Operations Management and Data Analytics Modelling

Operations Management and Data Analytics Modelling

Economic Crises Perspective

Edited by
Lalit Kumar Awasthi
Sushendra Kumar Misra
Dilbagh Panchal
Mohit Tyagi

CRC Press is an imprint of the
Taylor & Francis Group, an informa business

First edition published 2022
by CRC Press
6000 Broken Sound Parkway NW, Suite 300, Boca Raton, FL 33487-2742

and by CRC Press
2 Park Square, Milton Park, Abingdon, Oxon, OX14 4RN

First edition published by CRC Press 2022

CRC Press is an imprint of Taylor & Francis Group, LLC

ISBN: 978-0-367-75451-8 (hbk)
ISBN: 978-1-032-02059-4 (pbk)
ISBN: 978-1-003-18164-4 (ebk)

DOI: 10.1201/9781003181644

Typeset in Times
by Newgen Publishing UK

Dedication

This book is dedicated to the entire research group.

Contents

Preface

Operational analytics is a specific term for a type of business analytics which focuses on improving existing operations. This type of business analytics, like others, involves the use of various datamining and data aggregation tools to get more transparent information for business planning. In order to deal with the societal and economical challenges of the world, novel techniques/tools analysis output-based industrial practices play a vital role in improving the total production of industrial sectors. At present the world is facing a major problem related to business in different sectors due to the COVID-19 pandemic. The world economy is at its lowest level and it is going to be very difficult for all countries to emerge from this major problem. Therefore, to overcome this problem the role of operation management-based approaches increases manifold. For the advancement of operational management techniques and practices in different industrial sectors, it is essential to share innovative ideas and thoughts on a common platform where researchers across the globe can publish their innovative work to share novel ideas related to optimal solutions of various problems in the area of operation management, healthcare sector, supply chain management, reliability and maintenance management, quality management, financial management, product design and development, project management, service system and service management, energy and environment management, waste management, sustainable manufacturing and operations, systems design and performance measurement. It is a well-known fact that due to the high complexity of the issues related to society and the economy, interdisciplinary research is the key to future revolutions. From research funders to journal editors, policymakers to think tanks – all seem to agree that the future of research lies outside firm disciplinary boundaries. In such prevailing conditions, various working scenarios, conditions and strategies need to be optimized. Optimization is a multidisciplinary term and its essence can be inculcated in all domains of business, research and other associated working dynamics.

The aim of this book is bring together recent advances and trends in developing data-driven operation management-based methodologies, big data analysis, application of computers in industrial engineering, optimization techniques, development of decision support systems for Industrial operations, role of multiplecriteria decision-making (MCDM) approach in operation management, fuzzy set theory-based operation management modelling and Lean Six Sigma. The aim of this book is also to present the development of novel operation management-based frameworks and their application to different industrial sectors in the context of improving their productivity and sustainability under the current scenario. The scope of this book is not only limited to industrial sectors in the operation management field but it also covers other allied areas like financial management and human resource development, which are the thrust areas in the overall development of industrial profitability. It is also aimed to emphasize the effectiveness of proposed operation management approaches in comparison to the state-of-the-art existing approaches by means of illustrative examples and real-life industrial operational problems.

Contributors

Somesh Agarwal
Department of Industrial and Production Engineering
Dr. B.R. Ambedkar National Institute of Technology
Punjab, India

Mukesh Kumar Barua
Department of Management Studies
Indian Institute of Technology
Haridwar, India

Pushpendra S. Bharti
University School of Information, Communication & Technology
GGS Indraprastha University
New Delhi, India

G. S. Dangayach
Department of Mechanical Engineering
Malaviya National Institute of Technology
Jaipur, India

Mani Sankar Dasgupta
Department of Mechanical Engineering
Birla Institute of Technology and Science
Rajasthan, India

Joby G. David
Department of Mechanical Engineering
Amal Jyothi College of Engineering
Kerala, India

R. K. Garg
Department of Industrial and Production Engineering
Dr. B.R. Ambedkar National Institute of Technology
Punjab, India

Josy George
School of Mechanical Engineering
VIT Bhopal University
Bhopal (MP), India

Shrajal Gupta
Department of Mechanical Engineering
National Institute of Technology
Haryana, India

Shubham Gupta
Department of Instrumentation and Control Engineering
Dr. Bhimrao Ambedkar National Institute of Technology
Jalandhar, India

Sumit Gupta
Department of Mechanical Engineering
ASET, Amity University
Noida, India

Ajai Jain
Department of Mechanical Engineering
National Institute of Technology
Haryana, India

Chitranshu Khandelwal
Department of Management Studies
Indian Institute of Technology
Haridwar, India

Dinesh Khanduja
NIT
Kurukshetra, India

Anish Kumar
Department of Mechanical and Industrial Engineering,
Indian Institute of Technology
Uttrakhand, India

Neeraj Kumar
PKG College of Engineering and Technology
Haryana, India

Pradeep Kumar
Department of Mechanical and Industrial Engineering,
Indian Institute of Technology
Uttrakhand, India

Sourabh Kumar
Department of Management Studies
Indian Institute of Technology
Haridwar, India

Sachin Kumar Mangla
Jindal Global Business School
O P Jindal Global University
Haryana, India
Plymouth Business School
University of Plymouth
Plymouth, UK

Mebin Toms Mathew
Department of Mechanical Engineering
Amal Jyothi College of Engineering
Kerala, India

M. L. Meena
Department of Mechanical Engineering
Malaviya National Institute of Technology
Jaipur, India

Manoj Kumar Mishra
University Florida
Benin Campus, West Africa

Sambhu Nair V.S.
Department of Mechanical Engineering
Amal Jyothi College of Engineering
Kerala, India

Dilbagh Panchal
Department of Industrial and Production Engineering
Dr. B.R. Ambedkar National Institute of Technology
Punjab, India

Srikanta Routroy
Department of Mechanical Engineering
Birla Institute of Technology and Science
Rajasthan, India

Mijo P. Saji
Department of Mechanical Engineering
Amal Jyothi College of Engineering
Kerala, India

Prateek Saxena
Department of Industrial and Production Engineering
Dr. B.R. Ambedkar National Institute of Technology
Punjab, India

Amit Kumar Singh
Department of Instrumentation and Control Engineering
Dr. Bhimrao Ambedkar National Institute of Technology
Jalandhar, India

Shailender Singh
Department of Mechanical Engineering
Birla Institute of Technology and Science
Rajasthan, India

Imran Siraj
University School of Information, Communication & Technology
GGS Indraprastha University
New Delhi, India

Vikas Deepak Srivastava
JJT University Rajasthan
India

Dain D. Thomas
MRIIRS
Faridabad, India

Malhar Tidke
Department of Mechanical Engineering
Birla Institute of Technology and Science
Rajasthan, India

Mohit Tyagi
Department of Industrial and Production Engineering
Dr. B.R. Ambedkar National Institute of Technology
Punjab, India

Abi Varghese
Department of Mechanical Engineering
Amal Jyothi College of Engineering
Kerala, India

Karan Veer
Department of Instrumentation and Control Engineering
Dr. Bhimrao Ambedkar National Institute of Technology
Jalandhar, India

Keerti Veer
Department of Instrumentation and Control Engineering
Dr. Bhimrao Ambedkar National Institute of Technology
Jalandhar, India

Amit Vishwakarma
Department of Mechanical Engineering
Malaviya National Institute of Technology
Jaipur, India

J. Francis Xavier
School of Mechanical Engineering
VIT Bhopal University
Bhopal (MP), India

Organization of the Book

Chapter 1: This chapter throws light on the efficiency level of banking sector decision-making units by taking different components of the data envelopment analysis (DEA) technique with the objectivity of the Malmquist approach designed to measure total productivity of business entities engaged in similar kinds of decision-making activities. The core objective of this research study is to measure any change in efficiency level of these Indian banking sector organizations and their efficiency, along with their technical efficiency responsiveness to recent period paradigm tweaks in domestic and global events.

Chapter 2: This work proposes a multiple criteria decision-making (MCDM)-based framework to evaluate supplier selection by using an entropy weight method (EWM) for calculation of weightage of each criterion. Once the weightage is calculated the EWM is combined with a range of value (ROV) and evaluation based on distance from average solution (EDAS) method for supplier rank identification. Finally, the ranking performance of these methods is compared. A numerical example along with graphical illustrations is considered and comparison analysis is provided to test the feasibility of the proposed method. In the illustrative example a manufacturing firm is looking to select the most suitable supplier among ten suppliers based on four criteria: price/cost, service, quality and delivery, in which price/cost is non-beneficial and the attributes of the other criteria are beneficial.

Chapter 3: In this chapter, two tested, reliable and powerful techniques are synergistically combined to find the loss and optimize the process parameters by applying quality loss function (QLF) and response surface methodology (RSM). This study will further enhance the quality of the process and products.

Chapter 4: Coconut palm is also known as 'Kalpa Vriksha' since its different parts have multi-purpose utility. Coconuts are mainly used for their flesh, water and oil. Husking is one of the major post-harvesting operations associated with coconut. It is tedious in nature and mechanization of such a process demands standardization with respect to the physical attributes of coconut. The different physical attributes of coconut have a direct relationship with the mechanisms involved in the husking operation. Establishing a relationship between the physical attributes of coconuts and the designs of coconut husking mechanisms is vital for a successful husking process. This requires a study of the existing husking mechanisms and the various physical attributes of coconut. Hence there emerges the need to study the effects of the physical attributes of coconut on effective husking operations.

Chapter 5: In prosthetic arm control, to improve the robustness and reliability of different hand motions it is necessary for the arm to be controlled by highly accurate input or commands. To achieve high-accuracy input different types of classifier are used for pattern recognition. In this chapter the authors work with the widely used KNN (K-nearest neighbour) classifier and also determine time domain (TD) features. The output of the TD features works as the input of the classifier. As the input of the classifier increases, the accuracy of the classifier also increases. In this chapter the authors work with six different hand gestures/motions for pattern recognition. For all hand movements, a power spectrum density graph is also significantly analysed. The proposed techniques achieved 87% accuracy for different hand movements.

Chapter 6: This work proposes a responsive system for optimizing distance for waste electrical and electronic equipment (WEEE) collection vehicles using state-of-the-art genetic algorithms and ant colony optimization techniques. This will reduce the cost of collection for the formal sector, by reducing fuel consumption and service time and also providing a sustainable solution, reducing greenhouse gas emission. An Indian case scenario based on a service provider located in the city of Jaipur is developed. Certain assumptions are made, such as bulk consumers, urban households and an online platform. The model can be scaled up for the benefit of WEEE collection service providers in India with known arrangement of streets and buildings in a defined catchment area of a city.

Chapter 7: This study focuses on the necessity of clean energy sources and techniques by which we estimate solar energy/irradiance at any location. One way to estimate or forecast solar energy potential uses meteorological variables which are collected from various agencies in India. Accurate forecasting of solar energy depends on the soft computing technique used. With the help of a soft computing technique, we can rank the parameter responsible for solar energy potential/irradiance as well as correlations associated among the parameters. Various soft computing techniques are discussed in this chapter with their relative advantages and disadvantages.

Chapter 8: The aim of this chapter is to find out the important practices of sustainable product returns and recovery (SPRR) by conducting a survey in Indian textile industries. A survey instrument was developed on a five-point Likert scale. Respondents were asked about different issues related to SPRR. Responses were collected through an online survey using a convenient sampling method. Simple descriptions, correlations and *t*-tests were applied to the analysis of the survey responses. It was observed that reduced resource utilization (SPRR-1) and remanufacturing of returned products as the usable product (SPRR-5) were more practised by companies. SPRR issues were identified and empirically tested. The issues related to SPRR are highly significant for the clothing industry in India.

Chapter 9: Incorporating maintenance in scheduling problems has a significant impact on real-time scheduling problems. In this simulation study, reliability-centred preventive maintenance (RCPM) approach and job shop scheduling were integrated in a stochastic environment. The simulation model was constructed using Pro-Model simulation software. The manufacturing system consisted of six job types and ten different machines. The performance of the system was evaluated using four performance measures: makespan (C_{max}), mean flow time (MFT), mean tardiness (MT) and number of tardy jobs (NOTJ). Results demonstrated that lower levels of reliability, viz. 0.74, 0.78 and 0.82, are recommended for MFT and C_{max} performance measures. A 0.82 level of reliability is recommended for MT and NOTJ performance measures. Considering maintenance with scheduling, providing real-time scheduling work represents the novelty of the present study.

Chapter 10: This research work aims to develop a framework to access circular economy (CE) performance measures by taking the case of rubber waste from the automobile sector. Manufacturing of rubber and its end-of-life waste treatment emit a considerable amount of cancer-causing material and pollute the environment. CE's performance measures relevant to rubber industries are identified by reviewing the relevant published articles and interviewing industry personnel involved in a similar field. Through a questionnaire-based survey from the field expert team, data has been gathered; this is further analysed using a hybrid approach of analytic hierarchy process (AHP) and elimination and choice expressing reality (ELECTRE) approach. The proposed model delivers a pathway to policymakers/decision-makers for a deeper understanding of CE performance measurement, identifying the key performance measures for CE success, and a systemic transition towards CE.

Chapter 11: With the rise of accidental injury in India, it has become important to improve service quality in the healthcare industry, because any type of failure or error in the system significantly affects the safety of the patient as well as the goodwill of the hospital. An integrated failure mode and effect analysis (FMEA) and fuzzy logic rule-based inference model to study the different failure modes for risk assessment and to make corrective decisions were applied to this system. Under classical FMEA, risk priority numbers (RPNs) were tabulated using probability of occurrence (O), severity (S) and non-detection (D) values. In classical FMEA some causes of failure are difficult to distinguish in terms of their accurate priority. To overcome such limitations, a rule-based fuzzy logic based on the linguistic assessment of risk factors has been applied to compute fuzzy RPN. The ranking results were compared for effective decision making about the causes of the failure. From analysis of the results, cause C_3 was found to be the most critical one. The results were supplied to the hospital concerned for their implementation.

Chapter 12: The integration of Industry 4.0 (I4.0), an approach of technological progress and circular economy (CE) and an economic growth model that promotes cyclic resource usage, will aid in achieving organizational sustainability. Various factors help to drive I4.0 in the emerging markets for implementation of CE. Ten drivers were identified in this study based on an extensive literature review and expert opinion. Interpretative structural modelling (ISM) was used to form an interrelationship among the drivers, and MicMac was used to analyse the driving and dependence power of the drivers. The results indicate that management strategy and commitment are the most significant factors driving I4.0 for CE initiatives. This study gives managers and decision-makers information on I4.0 for successful and efficient implementation of CE in emerging markets.

Chapter 13: This chapter identifies key operational strategies to manage perishable food supply chains (PFSC). A mixed-methods approach was used based on a systematic literature review (SLR) and bibliometric analysis. This chapter further considers the case study of a dairy food supply chain (DFSC) in India. Dairy industry serves as the ideal case of an industry that strategically uses supply chain drivers to manage perishability, resulting in a high-performance supply chain. The analysis showed that first-mile technology application, effective cold chain management, stakeholder management and strict temperature control are some of the key aspects of managing perishability in PFSC. While cold chain serves as the backbone of the dairy supply chain, integration of farmers and supply chain responsiveness are other important aspects of managing PFSC. Managing perishability is a key factor for achieving global food safety, security and sustainable development goals. This chapter contributes towards sustainable food supply chains through successful management of perishability.

Chapter 14: This chapter discusses implemention of Lean Six Sigma (LSS) in relation to the environment, also known as Green Lean Six Sigma (GLSS). It provides a framework and the tools that can be used for implementation. The role of GLSS and LSS individually in this integration is also discussed.

Brief Introduction of Editors

Prof. Lalit Kumar Awasthi has been Director of the Dr B. R. Ambedkar National Institute of Technology (NIT), Jalandhar, India since 10 October 2016. He is a distinguished Professor of Computer Science and Engineering at NIT Hamirpur, Himachal, India. He has 30 years of teaching and research experience. He is a founder member of the Computer Science and Engineering Department at NIT Hamirpur. He has been instrumental in starting BTech, two MTech and PhD programmes at NIT Hamirpur. He has been a member of the Board of Governors of NIT Hamirpur for 2 years. His main portfolios at NIT Hamirpur include Head of the Computer Science and Engineering Department, Dean of students and alumni, Head of the Computer Centre and Hostel Administration. He has 150 research publications to his credit published in journals and conferences of international repute. He is a reviewer for the Elsevier journal *Ad Hoc Networks*, *IEEE Transactions on Computers* and many more. He is the Associate Editor of *International Journal of Computer Engineering and Information Technology.*

Dr Awasthi was instrumental in starting alumni chapters of NIT Hamirpur in Delhi, Calcutta, and California. He has guided ten PhD students and 20 MTech students in mobile computing information security, network security, cloud computing, vehicular ad hoc networks, grid computing and P2P networks. He remained Principal Investigator for five research projects funded by the Ministry of Human Resource Development (MHRD) and Himachal Pradesh (HP) Government. Further, he was consultant for a project on e-governance and automation for Assam University, Silchar.

Dr Awasthi is a senior member of the Institute of Electrical and Electronics Engineers (IEEE) and a distinguished member of the Association for Computing Machinery (ACM). He has travelled widely both in India and abroad and has visited the USA, the UK, Thailand, Nepal, Australia, Singapore, Italy and France.

He is the founding Director of Atal Bihari Vajpai Institute of Engineering and Technology, Paragati Nagar, and served the institute for 4 years. In addition, he holds the office of Director of Jawahar Lal Nehru Government Engineering College Sundernagar and Mahatama Gandhi Engineering College, Jeori, Camp Office Sundernagar. He remained Dean, Academics of Himachal Technical University, Hamirpur for a period of 2 years. He has held the additional charge of Director, NIT Delhi for more than 4 months. He has been a member of the Board of Studies of prestigious institutions such as Delhi Technological University (DTU), NIT Jalandhar, NIT Delhi, NIT Jaipur, Punjab University, Chandigarh and HP University, Shimla. He has acted as expert in faculty selections at different levels for prestigious institutions in India. He has delivered more than 80 expert lectures and invited talks in various conferences throughout the country as well as abroad.

Dr Sushendra Kumar Misra works at the Dr B. R. Ambedkar National Institute of Technology (NIT), Jalandhar as Registrar and Head of the Center of Continuing Education. He is an alumnus of the Indian Institute of Management (IIM) Bangalore and Syracuse University, New York, USA. He completed an MBA and PhD in Management. Dr Misra has worked as Director, Controller, Registrar and Dean at Punjab Technical University, Jalandhar. His main areas of work include educational planning and management, financial management, human resource management, public policy and skill development. Dr Misra has published many research papers and attended several conferences, seminars, workshops and training both in India and abroad. He has visited many countries such as Malaysia, Thailand, Australia, Slovenia, Singapore, Austria and the USA. Dr Misra is a life member of All India Management Association (AIMA), New Delhi; Indian Society for Technical Education (ISTE), New Delhi; Youth Hostels Association of India, New Delhi; Association of Indian Management Scholars International (AIMS), Houston, USA; Associate Fellow of World Business Institute, Australia; and Member of International Economic Development Research Center, Hong Kong.

Dr Dilbagh Panchal is currently working as Assistant Professor in the Department of Industrial and Production Engineering, Dr B. R. Ambedkar National Institute of Technology, Jalandhar.. He works in the area of reliability and maintenance engineering, fuzzy decision making and operation management. He obtained his Bachelor (Hons.) degree in Mechanical Engineering from Kurukshetra University, Kurukshetra in 2007 and Masters (gold medallist) in Manufacturing Technology in 2011 from Dr B. R. Ambedkar National Institute of Technology, Jalandhar. He has completed his PhD at the Indian Institute of Technology, Roorkee in 2016. Currently, three PhD scholars are working under him. Seven MTech dissertations have been guided by him and two are in progress. He has published 22 research papers in SCI/Scopus-indexed journals. Ten book chapters have also been published by him for reputed publishers. He has edited two books in his area of expertise and seven books are in progress. He has also attended ten international. He is a life member of various societies such as Operational Research Society of India (ORSI), Kolkata; Indian Society for Technical Education (ISTE), New Delhi; and Institute of Electrical and Electronics Engineers (IIIE), Mumbai. He is an associate editor of *International Journal of System Assurance and Engineering Management* (Springer). He is a regular reviewer for many journals, including *International Journal of Industrial and System Engineering* (Inderscience), *International Journal of Operational Research* (Inderscience), *OPSEARCH* (Springer) and *Applied Energy* (Elsevier).

Dr Mohit Tyagi is Assistant Professor in the Department of Industrial and Production Engineering at Dr B. R. Ambedkar National Institute of Technology (NIT), Jalandhar, India. He obtained his BTech (Hons) in Mechanical Engineering from Uttar Pradesh Technical University (UPTU) Lucknow in 2008 and MTech (gold medal) in Product Design and Development from Motilal Nehru National Institute of Technology (MNNIT) Allahabad in 2010. He completed his PhD at the Indian Institute of Technology, Roorkee in 2015. His areas of research are industrial engineering, supply chain management, corporate social responsibility, performance measurement system, data science and fuzzy inference systems. He has around 7 years of teaching and research experience. He has guided 18 postgraduate dissertations and 15 undergraduate projects. He is presently supervising two MTech and three PhD scholars. He has to his credit around 75 publications in international and national journals and proceedings of international conferences and book chapters by reputed publishers. Dr Mohit Tyagi is a reviewer of many international journal of repute, including *International Journal of Industrial Engineering: Theory, Application and Practices, Supply Chain Management: An International Journal, International Journal of Logistics System Management* and *Journal of Manufacturing Technology Management, Information Systems, Grey Systems: Theory and Applications*. He has organized three international conferences and webinars in collaboration with Indo-German Research Center, Department of Science and Technology (DST), Government of India. He has also organized seven Technical Education Quality Improvement Programme (TEQIP)-sponsored short-term courses/faculty development programmes in his area of expertise.

He has performed many academic and administrative responsibilities, including Assistant Public Relation Officer, NIT Jalandhar, Assistant Training Officer, Coordinator Academic (postgraduate), Warden of Boy's Hostels, member of Joint Admission Counseling (JAC-2015-Delhi), coordinator and committee member of various leading events (Utkansh, Bharat Dhwani, Hackathon (National technical Event) Induction cum Orientation Program for New Entrants, Technical Fest (TechNITi), Swach Bharat Mission Committee).

1 Measuring Banking Sector Efficiency
A Malmquist Approach

Manoj Kumar Mishra[1] *and*
Vikas Deepak Srivastava[2]

[1] Professor of Economics and Statistics, British American University Florida, Benin Campus, West Africa

[2] Assistant Professor of Economics and Management, JJT University Rajasthan, India

1.1 INTRODUCTION

Banks and finance are two essential ingredients for scripting the success growth story of any economy. They are complementary, forming the backbone of emerging economies. The Indian banking sector has been subjected to a slew of structural and institutional reforms, particularly in the neoliberal phase of banking reforms spanning that period, after the subprime financial crisis of international lending sectors. During the last decade, the role of Indian commercial banks has had a substantial vigorous uptick, so that India has become a leader of the fastest-growing economies of the world. The digital revolution has given banks a new flavor of digital world leaders, paving the way towards new dimensions of cost-effectiveness challenges and opportunities. The spillovers of the global slowdown have been seeping down through the Indian economy, hitting hard the profitability acumen of banking giants like State Bank of India (SBI), Punjab National Bank (PNB) and Bank of Baroda (BOB). This clearly supports the premise that there is a strong connection between banking instability and cyclical downtrend of the economy. Undeniably, it is difficult to imagine the prosperity of trade, industry, commerce and even the lifeline of all these major components of the Indian economy, the agriculture sector, without the custodian part played by the country's commercial banks.

In recent years incipient profitability and the stalking fear of insolvency of banking cheerleaders have prompted action-packed dramas like mergers, restructuring banking codes in the form of structural banking reforms, and so on. The continuum of disequilibrium of assets and liabilities triggered alarm and resulted in enactment of a stressed assets policy response in June 2019. Despite all these odds, running parallel to the banking system, the cashless economy dream of India is shaping up with a great deal of enthusiasm on the part of the banking

DOI: 10.1201/9781003181644-1

sector. This is evident from a steady rise in the volume of retail payments through electronic payment interfaces, from 92.6% in 2018 to 95% in 2019. Enhancing the Indian growth trajectory with sustainable expansion in income, consumption and investments are key points that the banking system must deal with in current global financial contagion channels. The core objectives of recent banking policy resolution are to tame the ramping-up bank stressed assets class by strategically ringfencing the banking system. All these policy responses have begun to show a somewhat silver lining in terms of quantitative reduction in gross bank non-performing assets (NPA), sliding from 11.2% to 9.1% in the year ending 2019. But, there is a long way is go on charting measures, raising and maintaining banking sector efficiency and productivity in compliance with financial prudence.

It is possible to evaluate the banking system of any economy by stepping into a detailed parametric review of bank balance sheets and have a complete in-depth diagnosis of the health of the banking system of any country. In an era of competitive efficiency, performance analysis has also gained momentum in banks, which are acting as business entities to level off their market competitors. Banks are now an indispensable part of the modern economic system, playing a vital role in channeling funds and financial assets either in cash or cashless transactions to drive different interconnected particles of the economy irrespective of their organizational setups and functions, like the nation feeder agriculture sector, trade, business, enterprises, export and import sectors. Thus, in the light of the significant and relentless role of the banking system, regular performance analysis is a matter of the utmost priority for any nation. Bank productivity measurement can be done via intensive analysis of different banking growth indicators by selective approaches evolved and refined from time to time. The litmus tests of the assets and liabilities portfolio of banks accompanied by diversified risks ratios are some of the most prevalent quality check norms. Banking productivity has a direct link with the allocative efficiency of the financial resources of citizens of a nation packed with banks and their optimum utilization to serve socio-economic interests to the best advantages and ultimately facilitate a sustainable rise in national welfare as a whole. Recent observations of growth trends of developing countries strikingly reflect a notable upswing in labor and capital productivity in line with the revival of the investment cycle. Here, the curiosity in research is how far the upturn in productivity of key growth drivers of the economy has masked a growth performance in the banking sector. So, analytical examination of the efficiency and productivity parameters of banks opens up headroom for an effective counter-cyclical response on the part of banks to avert unprecedented slippages and spillovers looming in the future.

Input–output and cost–benefit appraisals are conventional approaches to evaluate the performance of any organization. Gholambri (2014) has commented on the paramount importance of organizations to devise ways to remove inefficiency bugs by devoting resources to enhancing and maintaining high standards of efficiency in the economy during phases of economic stagnation. Banks are not merely bricks-ands-mortar business entities but they epitomize a country's economic growth by translating individual liquidity in the form of public deposits to productive utilization for the development of integral economic growth agents.

The scope of the current research lies in a fact-finding approach to gauge the performance of banks on quantity metrics from quality lens score and a comparative study of selected banking giants of the Indian economy. This will help to make a due balance between proficiency and efficiency norms and find out effective built-in stability for them.

In previous times, various approaches and analytical tools have been used to assess the performance parameters of banking players and to take a comparative benchmark on their inter-temporal performance on the grounds of organizational and managerial efficiency and service rendering scale. Different productivity indices have also been put in place to set up a point of reference for the measurement of organizations' operational performance scale. The Malmquist productivity index (MPI) is in wide use to measure total factor productivity engaged as inputs to derive the output of an organization.

1.2 REVIEW OF STUDIES AND RESEARCH TO MEASURE PERFORMANCE AND EFFICIENCY LEVELS OF BANKING ORGANIZATIONS

In an endeavor to ascertain the efficiency performance of financial institutions, a number of studies are centered on banking institutions, focusing on frontier analysis of productivity measurement of these institutions. With technological development, the data envelopment analysis (DEA) approach has been the most effective tool of operational research adopted to assess bank managerial efficiency. This technique is widely used among banking players for comparative assessment of different branches and to help locate efficiency gaps and resulting tweaks to improve the performance of less efficient branches.

Substantial literature on the performance analysis using the DAE technique of banking institutions is available. Some of these studies have been conducted at branch level for performance analysis whereas others are devoted to the entire banking institution. The studies of Paradi and Zhou (2014) ranging from 1985 to 2011 and Berger and Humphrey (1997) are path-breaking research in the context of DEA analysis of the banking sector; their objectivity sets a statistical scale to enable a level playing field. Fethi and Pasiouras (2010) have made attempts using DEA and frontier analysis reviews adopted across national banks in their operation research, aiming to evaluate the performance of branches. CCR (Charnes, Cooper and Rhodes) and BCC (Banker, Charnes and Cooper) are two models of linear programming as the DEA component to calculate the productivity of any organization. Apart from a 2×2 input–output model, the DEA technique has turned out to be an effective tool for measuring total productivity in multi-input and output models. This technique is based on the notion of operational decision-making units (DMUs) engaged in handling inputs and their respective outputs. The given parameter for all DMUs is that they possess a homogeneous nature from an operational point of view so that deduction of their performance analysis with artificial intelligence could enable a clear vision for comparison of their operational performance (Cooper et al., 2004). The performance of DMUs is measured on a set validated scale ranging from 0 to 1. Operational units approaching a value of 1 are deemed to be efficient, compared to those that score 0 or nearly 0. There are other significant scientific approaches to the framework of individual research in a range of performance measurement of banking and financial business complexes, namely that of Jahanshaloo et al. (2004).

Daraio and Simar (2007) stated that productivity and efficiency are child nodes of the input–output model of any business enterprise. Further, the distance between employed input and produced output with due technical diligence can be regarded as a measurement of the frontiers of productivity. In rather general jargon, the term productivity can be defined as ratio expressible between firm output and the combination of its factors to gain the desired level of output. Growth in productivity needs to run in tandem with growth of other technological parameters. But this does not happen all the time. Even though an organization may have the set-up and level of technology to survive and outperform other rivals in the business arena, their business and organizational skills may not be in harmony with socio-economic objectives. All these underlying facts may create conflicts of fine-tuned interest requiring a high degree of entrepreneurial capabilities and practical wisdom to reap the fruits of business profitability in harmony with social ideals.

1.3 OBJECTIVES OF THE STUDY

The objectives of this study were:

- to measure changes in the efficiency level of the banking sector for a reference period (2016–2020)
- to calculate total productive factor of selected banks in the reference period (2016–2020)
- to measure efficiency and technical changes of selected banks.

1.4 HYPOTHESIS OF THE STUDY

The set hypothesis of the problem at hand is as follows:

Null Hypothesis
There has been no significant change in the efficiency level of selected banks during the reference period.

Alternate Hypothesis
There has been a significant change in the efficiency level of selected banks during the reference period.

1.5 RESEARCH METHODOLOGY

In this research, the MPI was taken as the methodological approach to measure performance analysis of selected banks with a view towards evaluating the productivity level of these banking DMUs over the period from 2016 to 2020. The MPI technique is part of DEA in the form of input–output analysis of linear programming evolved and put into practice as a CCR model. This approach uses mathematical induction to quantitatively scale the various homogeneous DMUs over their performance frontiers. The MPI is further developed from the DEA technique created by Fare et al. (1992) in order to have a comparative performance analysis of different DMUs over periods of time. It uses weighting: all input variables are assigned weights as per their relative significance in output generation. In more general terms, it weights the input–output ratio of DMUs of a particular type of industry.Below linear programming models to measure the efficiency and technological efficiency change over periods are suggested.

$$\sum_{j}^{n} = \lambda X_{ij} \leq \Theta X_{i0} \quad I = 1,2,3,\ldots\ldots m$$

$$\sum_{j}^{n} = \lambda X_{ij} \leq \Theta X_{i0} \quad I = 1,2,3,\ldots\ldots s$$

$$\lambda \geq 0 \quad j = 1,2,3,\ldots\ldots\ldots n$$

This model points out the input-type model relating to the CCR approach of DEA. This model shows a finite optimal solution linearly related to inputs and output as $0 < \Theta < 1$.

The production and planning (PP) set is given below:

Minimization solution:

$$\sum_{j}^{n} = \lambda X_{ij} \leq \varnothing x_{i0} I = 1,2,3,\ldots\ldots m$$

$$\sum_{j}^{n} = \lambda X_{ij} \leq \varnothing x_{i0} \; I = 1,2,3,\ldots\ldots s$$

$$\sum^{n} j = \lambda j = 1$$

$$\lambda \geq 0 \quad j = 1,2,3,\ldots\ldots\ldots n$$

Thus, the DEA technique as a frontier approach to productivity measurement is a useful tool in the exploratory assessment of the efficiency of different DMUs.

In this study, the MPI total factor productivity measure is used in the comparative performance analysis of public sector banks selected as a sample DMU in the banking domain.

$$M(x^t, y^t, x^{t+1}, y^{t+1}) = [\frac{D^t}{D^t}(\frac{x^t}{x^t}\frac{x^{t+1}}{y^t}) \times \frac{D^t}{D^t}(\frac{x^t}{x^t}\frac{x^{t+1}}{y^t})]1/2$$

Further DEA analyses are given below.

1.5.1 Technological Change and Technical Efficiency Change

The equation above is the Malmquist total factor productivity approach devised to measure disaggregated efficiency change and technical efficiency change compounding total factor productivity change over time. x and y are input and output variables used for t and $t + 1$ periods and their respective distance on production possibility frontiers. Thus, M is a type of geometric mean of input–output analysis factor inputs and their relative output spread over different time dimensions. The change in M implies a corresponding change in level of productivity at DMU level. $\leq M \geq$ shows a value of M that floats between two closed intervals.

In this study, technological change index (TCI) and technical efficiency change index (TECI) were taken into consideration to examine the technological and relative efficiency shift in the banking sector. Both are embedded components of MPI. Using the Malmquist method and the DEA approach, a change in both constituents of productivity measurement was observed and calculated from the respective balance sheets and annual reports of selected banks; their efficiency movements were also captured.

Statistical Analysis Using R package

To capture the performance of selected banking sector DMUs and their descriptive analysis of other relevant performance indicators, the R statistical package was used (Table 1.1).

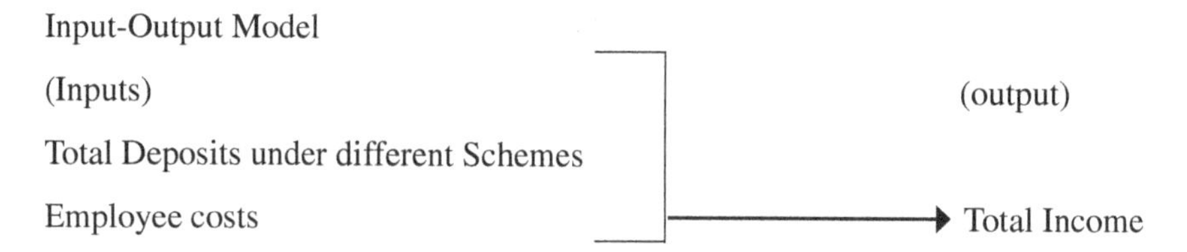

Total deposits and employee costs are key detriments of banking business and revenue generation.

1.5.2 Index for Measuring Total Factor Productivity Change and its Disaggregated Components

The MPI as data frontier analysis is applied to make a mathematical induction to locate gaps between two data points on the PP frontier. This can further be divided into other efficiency parameters such as input-biased technology change and output-biased technology change.

TABLE 1.1
Sample banks

S.N.	Decision-making unit	Type
1	State Bank of India	Public sector bank
2	Punjab National Bank	Public sector bank
3	Bank of Baroda	Public sector bank
4	Allahabad Bank	Public sector bank

1.6 RESULTS AND DISCUSSION

Figure 1.1 shows a trend of net profit/loss of respective banks during the study period (2006–2010). In 2018, the net loss for PNB was alarming, as it was downsizing in comparison to other banks.

From analysis of net NPA of selected banks for the study period, it is evident that mounting NPA is more challenging for SBI compared to other banks during the period taken as the reference year (Figure 1.2).

From Table 1.2, it is clear that there have been different average growth rates for the banks in question. In the case of SBI, for the period 2016–2020, 17.3% average growth rate is apparent compared to 14.8% for BOB, 6.2% for PNB and 2.6% for Allahabad Bank.

For remuneration and other provisions for employees working in these banking sector DMUs, a striking difference in terms of average growth rates for allocations and expenses incurred is also seen (Table 1.3). BOB ranks top among all other banks, registering 18.9% average growth for expenses such as employee costs.

Total income earned by all these banking sector DMUs from different branches operating across the country shows notable variation, including income received from all banking operations, 16.3% growth on average for a reference period in the case of BOB followed by SBI (12.3%), PNB (3.8%) and a surprisingly negative average growth rate of total income earned for Allahabad Bank (–3.8) (Table 1.4).

Table 1.5 reflects the summary statistics on input and output of DMUs operating in the banking sector. Generation of average income from various banking operation streams marks a quantitative difference for all sample banks for the period 2016–2020. SBI leads in the list with 250022.3 core income gained on

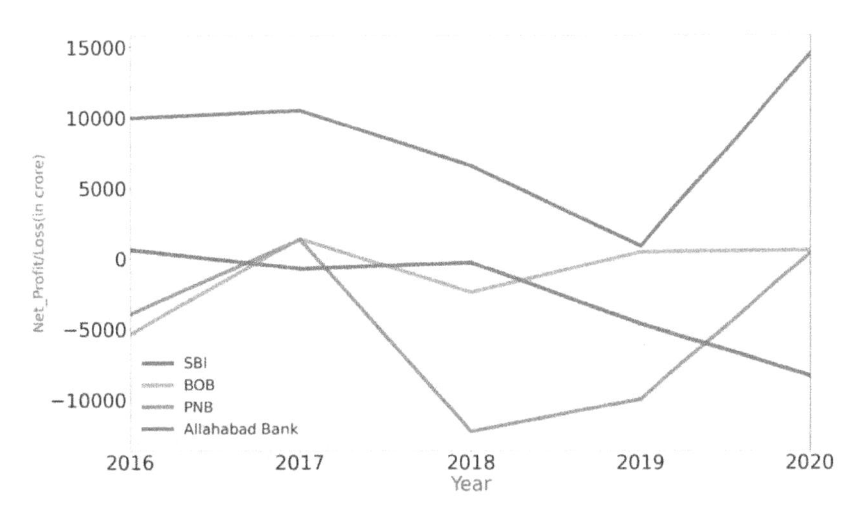

FIGURE 1.1 Net profit loss (2016–2020). SBI, State Bank of India; BOB, Bank of Baroda; PNB, Punjab National Bank.

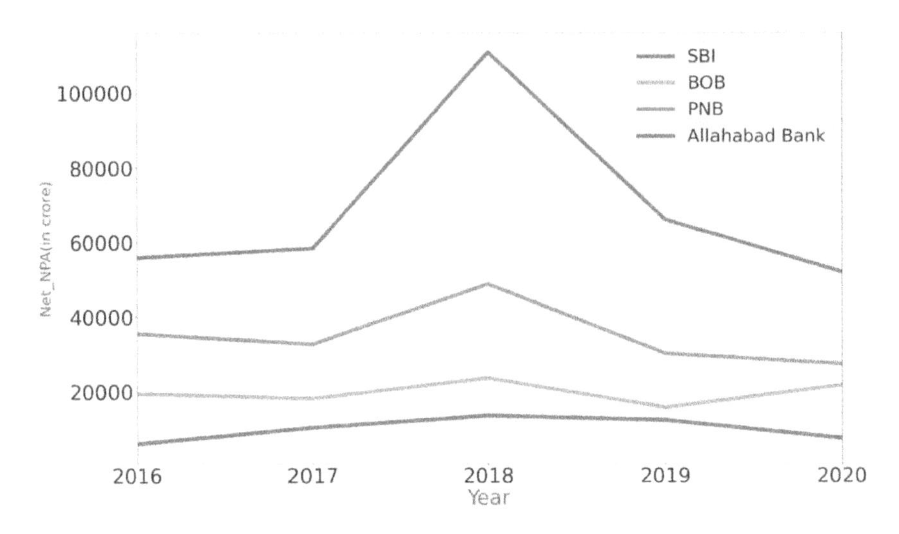

FIGURE 1.2 Net non-performing assets (NPA) (2016–2020). SBI, State Bank of India; BOB, Bank of Baroda; PNB, Punjab National Bank.

TABLE 1.2
Total deposits in core

Year	State Bank of India	Bank of Baroda	Punjab National Bank	Allahabad Bank
2016	1730722.44	574037.87	553051.13	193424.05
2017	2044751.39	601675.2	621704.00	200644.4
2018	2706343.00	591314.8	642226.2	201870.22
2019	2911386.00	638689.7	676030.1	213603.83
2020	3241621.00	945984.4	703846.3	214334.07
% growth (2016–2020)	17.3	14.8	6.2	2.6

Source: www.moneycontrol.com/india/financials/balance-sheet/

average as against average deposits (2526965 core) followed by 34310.29 core rupee spent on employee provisions.

A significant variation is too noted down here for change in MPI for all banks during reference years (Table 1.6). SBI maintained a consistent rise of MPI for successive years but, for other banks, MPI initially registered a decline before rising. Allahabad Bank shows a decline in MPI score for the year 2019–2020 before also gaining.

The efficiency change index for all banking sector DMUs is 1 for Allahabad Bank and BOB. Other banks achieved scores greater than 1, demonstrating greater efficiency attainment in earlier years but, in later periods of study, the efficiency change at DMUs is reflected by variations in scores (Table 1.7). BOB

TABLE 1.3
Total employee cost in core

Year	State Bank of India	Bank of Baroda	Punjab National Bank	Allahabad Bank
2016	25113.83	4978.03	6425.95	2307.2
2017	26489.28	4637.77	5420.72	2130.88
2018	33178.68	4606.87	9168.8	2285.94
2019	41054.71	5039.13	6963.16	2158.1
2020	45714.97	8769.52	6961.68	2529.87
% growth (2016–2020)	16.4	18.9	7.3	2.8

Source: www.moneycontrol.com/india/financials/balance-sheet/

TABLE 1.4
Total income in core

Year	State Bank of India	Bank of Baroda	Punjab National Bank	Allahabad Bank
2016	191843.67	49060.14	54301.37	21712.13
2017	210979.2	48957.99	56227.36	20795.07
2018	265100.00	50305.69	56876.64	20304.72
2019	279643.5	56065.1	58687.66	19051.05
% growth (2016–2020)	12.3	16.9	3.8	-3.8

Source: www.moneycontrol.com/india/financials/balance-sheet/

has maintained sustainability in efficiency across reference periods, in contrast to other banks that show a slight downward trend, again recapturing higher efficiency levels in subsequent years.

The technical efficiency scores achieved by all banking sector DMUs (>1 for BOB) shows a high level of technical efficiency shift for BOB (Table 1.8). For other banks, technical efficiency scores are also around 1, demonstrating better technical efficiency shifts; however, on an annual basis these scores reveal marked variations.

1.7 CONCLUSION

The research findings show the comparative differences in banking sector DMUs in performance level in terms of MPI, electronic commerce indicator (ECI), and TCI scores that further show an input–output matrix employed to achieve a set output. It is evident from this research study conducted for the purpose of assessing different levels of efficiency at operational level of these banking giants

TABLE 1.5
Descriptive statistics of banks for period 2016–2020

Bank	Total income (in core)				Total deposits (in core)				Total employee costs (in core)			
	Mean	Min.	Max.	sd	Mean	Min.	Max.	sd	Mean	Min.	Max.	sd
Allahabad Bank	20085.49	18564.5	21712.13	1282.673	204775.3	193424.1	214334.1	8995.384	2282.398	2130.88	2529.87	158.3333
BOB	58137.98	48957.99	86300.98	16011.52	670340.4	574037.9	945984.4	155897	5606.264	4606.87	8769.52	1778.994
PNB	57833.44	54301.37	63074.16	3322.85	639371.6	553051.1	703846.3	57602.85	6988.062	5420.72	9168.8	1371.994
SBI	250022.3	191843.7	302545.1	46830.85	2526965	1730722	3241621	623943.3	34310.29	25113.83	45714.97	8980.222

BOB, Bank of Baroda; PNB, Punjab National Bank; SBI, State Bank of India.

Source: author calculations from data derived from respective bank balance sheets for selected years.

TABLE 1.6
Malmquist productivity index for selected banks for period 2016–2020

Bank	2016–17	2017–18	2018–19	2019–20	Average
Allahabad Bank	0.9710	0.9554	0.9194	0.9553	0.95
Bank of Baroda	1.0621	1.03524	1.0227	0.9488	1.01
Punjab National Bank	1.1146	0.8611	1.0304	1.0363	1.01
State Bank of India	0.9308	0.9493	0.9679	0.9716	0.95
Average	1.01	0.95	0.98	0.97	

Source: author calculation using R package.

TABLE 1.7
Efficiency change for selected banks for period 2016–2020

Bank	2016–17	2017–18	2018–19	2019–20	Average
Allahabad Bank	1	1	0.9830	0.9542	0.98
Bank of Baroda	1	1	1	1	1
Punjab National Bank	1.1173	0.8804	1.0835	1.0265	1.02
State Bank of India	1.0081	0.9782	1.0268	1	1.00
Average	1.03	0.96	1.02	0.99	

Source: author calculation using R package.

TABLE 1.8
Technology change for selected banks for period 2016–2020

Bank	2016–17	2017–18	2018–19	2019–20	Average
Allahabad Bank	0.9710	0.9554	0.9353	1.001	0.96
Bank of Baroda	1.0621	1.0352	1.022	0.9488	1.01
Punjab National Bank	0.9975	0.9780	0.9509	1.0095	0.98
State Bank of India	0.9232	0.9704	0.9426	0.9716	0.95
Average	0.98	0.98	0.96	0.98	

Source: author calculation using R package.

that human resources play a dominant role in determining the efficiency level of any organization, as well as technical progress and its rational use with human capital. Even those periods studied were marked by twists and turns and global volatility. The MPI score for SBI is worth mentioning due to its consistency. That again reveals a prudent translation of its resources and their efficient utilization in contrast to other banking counterparts. The results on technical and efficiency scores show that, in banking, the use of technology is desirable in rational management, combined with other resources to the best possible advantage, creating a positive shift in technical improvisation in efficiency level. The qualitative combination of the workforce of any organization plays a significant game-changing role in setting the parameters of productivity on a global competitiveness platform.

Further scope of this research is to use wider banking organizations operating in money and finance markets and for longer periods, together with greater inclusion of input and output that significantly influences the productivity and operational efficiency of these DMUs.

REFERENCES

Berger, A.N., Humphrey, D.B., 1997. Efficiency of financial institutions: international survey and directions for future. European Journal of Operational Research 98, 175–212.

Cooper, W.W., Deng, H., Huang, Z., Li, S. X. 2004. Chance constrained programming approaches to congestion in stochastic data envelopment analysis. European Journal of Operational Research 155(2), 487–501.

Daraio, C., Simar, L., 2007. Advanced Robust and Nonparametric Methods in Efficiency Analysis: Methodology and Applications. New York: Springer.

Fare, R., Grosskopf, S., Lindgren, B., Roos, P., 1992. Productivity change in Swedish pharmacies 1980–1989. A non parametric Malmquist approach. Journal of Productivity Analysis 3, 85–102.

Fethi, M.D., Pasiouras, F., 2010. Assessing bank efficiency and performance with operational research and artificial intelligence techniques: a survey. European Journal of Operational Research 204, 189–198.

Gholam Abri, A.A., 2014. Evaluating the efficiency of social security in Isfahan Province. Economic Modelling 8, 83–99.

Jahanshahloo, G.R., Hosseinzadeh Lotfi, F., Shoja, N., Tohidi, G., Razavian, S., 2004. Ranking using L1 norm in data envelopment analysis. Applied Mathematics and Computational 153, 215–224.

Paradi, J.C., Zhu, H., 2014. A survey on bank branch efficiency and performance research with data envelopment analysis. Omega 41, 61–79.

2 A Hybrid MCDM Model Combining Entropy Weight Method with Range of Value (ROV) Method and Evaluation Based on Distance from Average Solution (EDAS) Method for Supplier Selection in Supply Chain Management

Josy George[1] *and J. Francis Xavier*[2]

[1] PhD Scholar, School of Mechanical Engineering, VIT Bhopal University, Bhopal (MP), India

[2] Professor, School of Mechanical Engineering, VIT Bhopal University, Bhopal (MP), India

DOI: 10.1201/9781003181644-2

2.1 INTRODUCTION

As supplier selection and evaluation is the basic exercise of the purchase department of any organization, it can be said that supplier selection is the one of the most important links of supply chain management. It is very important for the purchase manager or department to select a suitable supplier; otherwise, it is not possible for organizations to sustain product or service quality based on benchmarks. There are many methods to select the best among a group of suppliers. Multiple criteria decision-making (MCDM) techniques are used to create appropriate solutions for complex problems dealing with different criteria and alternatives. Basically, MCDM works with the priorities of the criteria related to the problem objectives, with the help of defined mathematic algorithms for each and every MCDM technique. After the algorithms have been applied the results help decision makers to make a judgment.

2.2 LITERATURE

Supplier selection methods are the various approaches to find the most suitable supplier for business purposes. Many selection methods have been discussed by researchers. There is a rapid increase in work aggregating sustainability using a variety of MCDM. MCDM techniques are powerful tools used to evaluate and select related problems. In this chapter a new integrated model was discussed and compared for supplier selection based on different critera and alternates. In the present study, weights of the criteria were calculated by the entropy weight method (EWM) and then the performance of different alternatives were ranked by evaluation based on distance from average solution (EDAS) and range of value (ROV).

2.2.1 Entropy Weight Method

The concept of information entropy was first introduced by Claude E. Shannon. Nowadays, it has been widely used in engineering, economy, and finance. Information entropy is the measurement of the degree of disorder of a system. It can measure the amount of useful information with the data provided. When the difference between the value of the evaluating objects on the same indicator is high while the entropy is small, this indicator provides more useful information, and the weight of this indicator should be set correspondingly high. On the other hand, if the difference is smaller and the entropy is higher, the relative weight would be smaller. Hence, the entropy theory is an objective way for weight determination [1, 2].

According to Shannon, entropy can be used to ascertain the degree of disorder and its utility in system information. The smaller the entropy value, the smaller the degree of disorder in the system. The index weight is determined by the amount of information based on EWM, which is an objective fixed-weight method [3].

EWM includes the following five steps [4, 5]:

Step 1: Construction of a decision matrix (X). A set of alternatives ($A = \{A_i, i = 1, 2 \ldots., n\}$) is to be compared to a set of criteria ($C = \{C_j, j = 1, 2 \ldots, m\}$). Therefore, an $n*m$ performance matrix (the decision matrix; X) can be obtained as:

$$X = \begin{bmatrix} x_{12} & x_{12} & \cdots & x_{1n} \\ x_{21} & x_{22} & \cdots & x_{2n} \\ \cdots & \cdots & \cdots & \cdots \\ \cdots & \cdots & \cdots & \cdots \\ \cdots & \cdots & \cdots & \cdots \\ x_{m1} & x_{m2} & \cdots & x_{mn} \end{bmatrix} \quad (1)$$

where x_{ij} is a crisp value indicating the performance rating of each alternative A_i with regard to each criterion C_j.

Step 2: To ascertain objective weights by the entropy measure, the decision matrix in Eq. (1) needs to be normalized for each criterion C_j ($j = 1,2 \ldots m$)

$$p_{ij} = \frac{x_{ij}}{\sum_{i=1}^{m} x_{ij}} \quad (2)$$

The normalized decision matrix is obtained as a result of the process:

$$P = \begin{bmatrix} p_{12} & p_{12} & \cdots & p_{1n} \\ p_{21} & p_{22} & \cdots & p_{2n} \\ \cdots & \cdots & \cdots & \cdots \\ \cdots & \cdots & \cdots & \cdots \\ \cdots & \cdots & \cdots & \cdots \\ p_{m1} & p_{m2} & \cdots & p_{mn} \end{bmatrix} \quad (3)$$

Step 3: Calculate the entropy measure of every index using the following equation:

$$e_j = -k \sum_{i=1}^{m} p_{ij} . ln\, p_{ij} \quad (4)$$

where $k = \frac{1}{ln(n)}$ is a constant which guarantees $0 < e_j < 1$

Step 4: The degree of divergence (d_j) of the average intrinsic information contained by each criterion C_j ($j = 1, 2 \ldots, m$) can be calculated as:

$$d_j = 1 - e_j \tag{5}$$

Step 5: The objective weight for each criterion C_j (j = 1, 2 ..., m) is thus given by:

$$w_j = \frac{1 - e_j}{\sum_{j=1}^{n} 1 - e_j} \tag{6}$$

2.2.2 Evaluation Based on Distance from Average Solution Method

EDAS was introduced to the world by Mehdi Keshavarz Ghorabaee; in EDAS the distances in both positive and negative directions are calculated from the average solution separately and according to the beneficial or non-beneficial criteria chosen [6]. The desirability of alternatives in this method is determined based on distances from an average solution. Here we have two measures dealing with the desirability of the alternatives. The first is the positive distance from average (PDA), and the second is the negative distance from average (NDA). These measures show the difference between each solution (alternative) and the average solution. The evaluation of the alternatives is made according to higher values of PDA and lower values of NDA. Higher values of PDA and/or lower values of NDA indicate that the solution (alternative) is better than the average solution [7, 8].

The main steps of the EDAS method are as follows:

Step 1. Select the most important criteria that describe alternatives.
Step 2. Construct the decision-making matrix X, as follows:

$$X = \left[x_{ij} \right]_{n*m} = \begin{bmatrix} x_{11} & x_{12} & \cdots & x_{1n} \\ x_{21} & x_{22} & \cdots & x_{2n} \\ \cdot & \cdot & \cdots & \cdot \\ \cdot & \cdot & \cdots & \cdot \\ \cdot & \cdot & \cdots & \cdot \\ x_{m1} & x_{m2} & \cdots & x_{mn} \end{bmatrix} \tag{7}$$

where x_{ij} ($x_{ij} \geq 0$) denotes the performance value of *i-th* alternative on *j-th* criterion ($i \in \{1,2, ..., n\}$ and $j \in \{1,2, ..., m\}$).

Step 3. Determine the average solution according to all criteria, shown as follows:

$$AV = \left[AV_j \right]_{1*m} \tag{8}$$

where:

$$AV_j = \frac{\sum_{i=1}^{n} X_{ij}}{n} \tag{9}$$

Step 4: Calculate the PDA and NDA matrixes according to the type of criteria (benefit and cost), as follows:

$$PDA = \left[PDA_{ij} \right]_{n*m} \tag{10}$$

$$NDA = \left[NDA_{ij} \right]_{n*m} \tag{11}$$

If *j-th* criterion is beneficial:

$$PDA_{ij} = \frac{max\left(0, \left(X_{ij} - AV_j\right)\right)}{AV_j} \tag{12}$$

$$NDA_{ij} = \frac{max\left(0, \left(AV_j - X_{ij}\right)\right)}{AV_j} \tag{13}$$

And if the *j-th* criterion is non-beneficial:

$$PDA_{ij} = \frac{max\left(0, \left(AV_j - X_{ij}\right)\right)}{AV_j} \tag{14}$$

$$NDA_{ij} = \frac{max\left(0, \left(X_{ij} - AV_j\right)\right)}{AV_j} \tag{15}$$

where PDA_{ij} and NDA_{ij} denote the positive and negative distance of *i-th* alternative from average solution in terms of *j-th* criterion, respectively.

Step 5. Determine the weighted sum of PDA and NDA for all alternatives, as follows:

$$SP_i = \sum_{j=1}^{m} w_j PDA_{ij} \tag{16}$$

$$NP_i = \sum_{j=1}^{m} w_j NDA_{ij} \tag{17}$$

where w_j is the weight of *j-th* criterion.

Step 6. Normalize the values of weighted sum of positive distance (SP) and weighted sum of negative distance (SN) for all alternatives, as follows:

$$NSP_i = \frac{SP_i}{max_i(SP_i)} \tag{18}$$

$$NSN_i = 1 - \frac{SN_i}{max_i(SN_i)} \tag{19}$$

Step 7. Calculate the appraisal score (AS) for all the alternatives, as follows:

$$AS_i = \frac{1}{2}(NSP_i + NSN_i) \tag{20}$$

where $0 \le AS_i \le 1$.

2.2.3 Range of Value Method

The ROV method is used to rank alternatives [9]. This method only requires the ordinal specification of important criteria from a decision-maker. Thus, in situations where decision-makers are facing problems in supplying quantitative weights, the ROV method can be particularly useful [10].

The application of any MCDM method to solve a decision-making problem usually involves three main steps: (1) determination of the relevant conflicting criteria and feasible alternatives; (2) measurement of the relative importance of the considered criteria and impact of the alternatives on those criteria; and (3) determination of the performance measures of the alternatives for ranking [11]. The procedure of the application of the multi-objective optimization on the basis of ratio analysis (MOORA) method is simple and consists of the following steps:

Step 1: The ROV method starts with setting the goals and identification of the relevant criteria for evaluating available alternatives.

Step 2: Based on the information available about the alternatives, a decision-making matrix or decision table is set. Each row refers to an alternative and each column to one criterion. The initial decision matrix, X, is:

$$X = [x_{ij}] = \begin{bmatrix} x_{11} & x_{12} & \cdots & x_{1n} \\ x_{21} & x_{22} & \cdots & x_{2n} \\ . & . & \cdots & . \\ . & . & \cdots & . \\ . & . & \cdots & . \\ x_{m1} & x_{m2} & \cdots & x_{mn} \end{bmatrix} \tag{21}$$

where x_{ij} is the performance measure of *i-th* alternative concerning *j-th* criterion, m is the number of alternatives, and n is the number of criteria.

Step 3. Performance measures of alternatives are normalized, defining values $\bar{x}_{ij}$ of normalized decision-making matrix $\bar{X}$.

$$X = \left[\bar{x}_{ij}\right] = \begin{bmatrix} \bar{x}_{11} & \bar{x}_{12} & \cdots & \bar{x}_{1n} \\ \bar{x}_{21} & \bar{x}_{21} & \cdots & \bar{x}_{2n} \\ . & . & \cdots & . \\ . & . & \cdots & . \\ . & . & \cdots & . \\ \bar{x}_{m1} & \bar{x}_{m1} & \cdots & \bar{x}_{mn} \end{bmatrix} \tag{22}$$

For beneficial criteria, whose preferable values are maxima, normalization is done using the linear transformation [10]:

$$\bar{x}_{ij} = \frac{x_{ij} - min_{i-1}^{m}\left(x_{ij}\right)}{max_{i-1}^{m}\left(x_{ij}\right) - min_{i-1}^{m}\left(x_{ij}\right)} \tag{23}$$

For non-beneficial criteria, whose preferable values are minima, normalization is done by:

$$\bar{x}_{ij} = \frac{max_{i-1}^{m}\left(x_{ij}\right) - x_{ij}}{max_{i-1}^{m}\left(x_{ij}\right) - min_{i-1}^{m}\left(x_{ij}\right)} \tag{24}$$

Step 4. Application of the ROV method involves calculation of the best and worst utility for each alternative. This is achieved by maximizing and minimizing a utility function. For a linear additive model, the best utility (u_i^+) and the worst utility (u_i^-) of *i-th* alternative are obtained using the following equations [10, 12]:

$$\text{Maximize: } u_i^+ = \sum_{j=1}^{n} \bar{x}_{ij}.w_j \tag{25}$$

$$\text{Minimize: } u_i^- = \sum_{j=1}^{n} \bar{x}_{ij}.w_j \tag{26}$$

where w_j ($j = 1 \ldots, n$) are criteria weights which satisfy $\sum_{j=1}^{n} w_j = 1$ and $w_j \geq 0$

If $u_i^- > u_i^+$ then alternative i outperforms alternative i regardless of the actual quantitative weights. If it is not possible to differentiate the options on this basis then a score (enabling subsequent ranking) can be attained from the mid-point, which can be calculated as [10, 12]:

$$u_i = \frac{u_i^- + u_i^+}{2} \tag{27}$$

Step 5. In this final step, the complete ordinal ranking of the alternatives is obtained based on u_i values. Thus, the best alternative has the highest u_i value and the worst alternative has the lowest u_i value.

2.3 ILLUSTRATIVE EXAMPLE

In this section, to verify the methodology using a simulated numerical example, we report a case study where the management of a manufacturing company wants to choose their best suppliers based on proposed methodology.

In a given problem, a textile firm wants to choose their best supplier to provide machinery parts for their maintenance division. In the selection process, ten suppliers (A1–A10) were considered as well as four selection criteria: that are product quality (PQ), price (PR), service quality (SQ), and delivery time (DT). Among these four criteria, PQ, SQ, and DT are beneficial (i.e., higher values are desirable), whereas one criterion (PR) is non-beneficial (Table 2.1).

2.3.1 Implementation of EWM Method

EWM is a commonly used weighting method that measures value dispersion in decision-making. In the given problem in the first stage the EWM is used to calculate the weight of the given criteria. After implemention of the mathematic

TABLE 2.1
Decision matrix

Alternatives	Product quality	Product cost	Delivery time	After-service
A1	12	90	92	205
A2	9	75	88	170
A3	10	60	91	190
A4	15	75	83	175
A5	12	70	89	165
A6	15	60	92	215
A7	11	65	91	160
A8	10	80	75	180
A9	14	50	91	200
A10	13	90	86	195
Criteria type	Maximum	Minimum	Maximum	Maximum

TABLE 2.2
Conversion matrix with weight

Alternatives	Product quality	Product cost	Delivery time	After-service
A1	0.0992	0.1259	0.1048	0.1105
A2	0.0744	0.1049	0.1002	0.0916
A3	0.0826	0.0839	0.1036	0.1024
A4	0.124	0.1049	0.0945	0.0943
A5	0.0992	0.0979	0.1014	0.0889
A6	0.124	0.0839	0.1048	0.1159
A7	0.0909	0.0909	0.1036	0.0863
A8	0.0826	0.1119	0.0854	0.097
A9	0.1157	0.0699	0.1036	0.1078
A10	0.1074	0.1259	0.0979	0.1051
Weightage (W_j)	0.3952	0.4326	0.0486	0.1236

TABLE 2.3
Normalized values of normalized sum of positive distance (NSP_i) and normalized sum of negative distance (NSN_i) with supplier ranks

Alternatives	NSP_i	NSN_i	Appraisal scores (Asi)	Rank
A1	0.0752	0.8848	0.4800	8
A2	0.0005	0.8672	0.4339	10
A3	0.3652	0.9314	0.6483	4
A4	0.4653	0.9692	0.7172	3
A5	0.0478	0.9831	0.5155	7
A6	0.9150	1.0000	0.9575	2
A7	0.2019	0.9471	0.5745	5
A8	0.0000	0.8692	0.4346	9
A9	1.0000	1.0000	1.0000	1
A10	0.1755	0.8871	0.5313	6

algorithms the weight is calculated (Table 2.2). The results are calculated with the help of MS Excel.

2.3.2 Implementation of EDAS Method for Ranking of Supplier

The mathematical procedure for the EDAS method for a decision-making problem with four criteria and ten alternatives is obtained with the help of the procedure defined in section 2.2 (Table 2.3).

2.3.3 Implementation of ROV Method for Ranking of Supplier

The ROV method is used to ranking the alternatives. Firstly, the normalized decision matrix is obtained following the equations given and then the best and worst

TABLE 2.4
Final ranking for all alternatives

Alternatives	u_i^+	u_i^-	u_i	Rank
A1	0.0000	0.3473	0.1737	8
A2	0.1622	0.0596	0.1109	9
A3	0.3245	0.1790	0.2517	4
A4	0.1622	0.4518	0.3070	3
A5	0.2163	0.2489	0.2326	5
A6	0.3245	0.5674	0.4459	2
A7	0.2704	0.1775	0.2239	6
A8	0.1082	0.1108	0.1095	10
A9	0.4326	0.4650	0.4488	1
A10	0.0000	0.3736	0.1868	7

utility functions for each alternative are calculated using the equations mentioned in section 2.3. The criteria weights derived from the entropy method are used to make these calculations. Finally, the u_i values of all alternatives with respect to the criteria considered are estimated. Table 2.4 shows the results of the ROV method according to which complete ranking of the alternatives is obtained.

2.4 RESULTS

Supplier evaluation and selection have been identified as important problems which can affect the efficiency of a supply chain. In this chapter, we have proposed a hybrid model to combine EWM with EDAS and ROV. The results of this study are illustrated with a numerical example which successfully demonstrates supplier selection. After implementation of EDAS and ROV methods, we have the ranking of alternatives: A9>A6>A4>A3>A7>A10>A5>A1>A8>A2 and A9>A6>A4>A3>A5>A7>A10>A1>A2>A8. Both methods identified that A9 is the most suitable supplier for the organization, as according to the results obtained with both methods up to the fourth alternative the results are the same. However, when moving forward to the next alternative and further, the difference in the methods becomes clear; there were considerable disparities in outcomes and a great deal of uncertainty as regards other suppliers. Although the outputs obtained explicitly clarified that there is a potential risk in adhering to the results of a single MCDM technique, based on the result both EDAS and ROV methods integrated with EWM are effective in solving supplier selection problems (Figure 2.1).

2.5 CONCLUSION

MCDM is an effective tool used to solve complex selection issues, including multiple criteria and options, especially for qualitative variables. Both methods described are based on an evaluation matrix and they can simultaneously consider

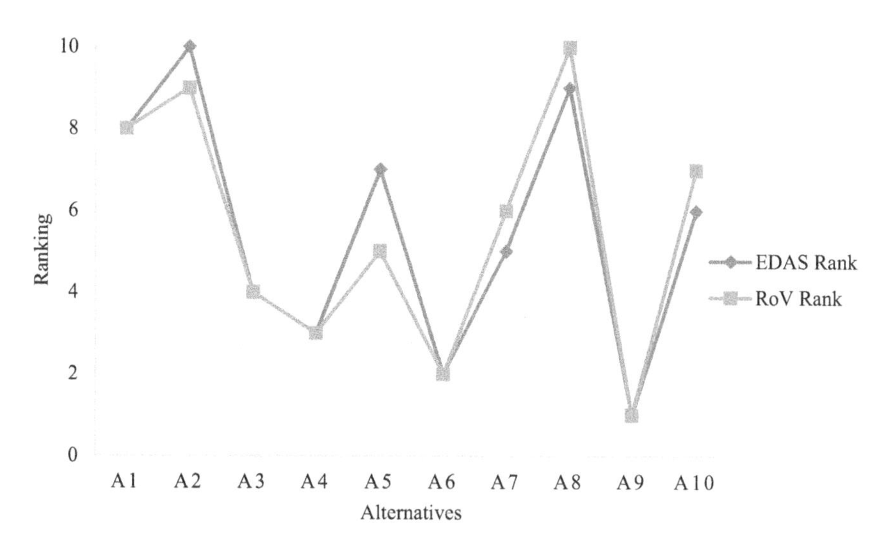

FIGURE 2.1 Comparison of results of evaluation based on distance from average solution (EDAS) and range of value (RoV) methods.

any number of criteria and alternatives. So complex decision problems can be organized and solved in a consistent manner. They handle the beneficial and non-beneficial criteria in the problem separately. They follow a simple computational procedure and can be used with MS Excel. These methods are suitable for the analysis of alternative performances with respect to various conflicting criteria, both qualitative and quantitative. This chapter shows that EWM integrated with EDAS and ROV performed efficiently for the supplier selection problem. In future studies, a proposed combined approach may also be applied to other selection problems of any organization. The number of evaluation criteria and alternatives may be changed according to the needs of the organization. The weights of the criteria may be derived from different weighting methods. The ranking of the alternatives may be performed with other MCDM methods and the results obtained compared.

REFERENCES

1. Işık, A. T., & Adalı, E. A. (2017). The decision-making approach based on the combination of entropy and rov methods for the apple selection problem. *European Journal of Interdisciplinary Studies*, *3*(3), 80–86. https://doi.org/10.26417/ejis.v8i1.p81-6
2. Zou, Z. H., Yi, Y., & Sun, J. N. (2006). Entropy method for determination of weight of evaluating indicators in fuzzy synthetic evaluation for water quality assessment. Journal of Environmental Sciences, 18(5), 1020–1023.
3. Li, X., Wang, K., Liu, L., Xin, J., Yang, H., & Gao, C. (2011). Application of the entropy weight and TOPSIS method in safety evaluation of coal mines. Procedia Engineering, 26, 2085–2091.

4. Deng, H., Yeh, C. H., & Willis, R. J. (2000). Inter-company comparison using modified TOPSIS with objective weights. Computers & Operations Research, 27(10), 963–973.
5. Shemshadi, A., Shirazi, H., Toreihi, M., & Tarokh, M. J. (2011). A fuzzy VIKOR method for supplier selection based on entropy measure for objective weighting. Expert Systems with Applications, 38(10), 12160–12167.
6. Ghorabaee, M. K., Zavadskas, E. K., Olfat, L., & Turskis, Z. (2015). Multi-criteria inventory classification using a new method of evaluation based on distance from average solution (EDAS). Informatica, 26(3), 435–451, doi: 10.15388/Informatica.2015.57. 34.
7. Ghorabaee, M. K., Zavadskas, E. K., Amiri, M., & Turskis, Z. (2016). Extended EDAS method for fuzzy multi-criteria decision-making: an application to supplier selection. International Journal of Computers Communications & Control, 11(3), 358–371, doi: 10.15837/ijccc.2016.3.2557.
8. Keshavarz Ghorabaee, M., Amiri, M., Zavadskas, E. K., Turskis, Z., & Antucheviciene, J. (2017). A new multi-criteria model based on interval type-2 fuzzy sets and EDAS method for supplier evaluation and order allocation with environmental considerations. Computers and Industrial Engineering, 112, 156–174. https://doi.org/10.1016/j.cie.2017.08.017
9. Yakowitz, D. S., Lane, L. J., & Szidarovszky, F. (1993). Multi-attribute decision making: dominance with respect to an importance order of the attributes. Applied Mathematics and Computation, 54(2), 167–181.
10. Hajkowicz, S., & Higgins, A. (2008). A comparison of multiple criteria analysis techniques for water resource management. European Journal of Operational Research, 184(1), 255–265.
11. Chakraborty, S., & Chatterjee, P. (2013). Selection of materials using multi-criteria decision-making methods with minimum data. Decision Science Letters, 2(3), 135–148.
12. Athawale, V. M., & Chakraborty, S. (2011). A comparative study on the ranking performance of some multi-criteria decision-making methods for industrial robot selection. International Journal of Industrial Engineering Computations, 2(4), 831–850.

3 Quality Loss Function Deployment in Fused Deposition Modelling

Imran Siraj and Pushpendra S. Bharti
University School of Information, Communication & Technology, GGS IndraPrastha University, New Delhi, India

3.1 INTRODUCTION

Fused deposition modelling (FDM) technologies are technologies that are rapidly taking over from conventional manufacturing to new important areas of modelling and fabrication of unimagined sectors (Sun et al., 2013). These technologies are swiftly gaining pace in application in the manufacture of conventional machines and automobile parts, production of highly precise components of instrumentation and control industry, research and scientific exploration of astronomy and space engineering (Du et al., 2018). Another swift area of penetration is re-engineering of bone and organ implants and regenerative tissue systems (Zhu et al., 2016). Since the human body cannot afford the correction of another implantat, surgery involving such organs needs to be a complete success on the first attempt.

This requires a lot of careful research and expertise in the area of FDM (Zheng et al., 2016). A high degree of accuracy is the need of the hour, compelling researchers to find the quality in the process and achieve the target value (Macdonald et al., 2014). Frequently used materials are thermoplastic polylactic acid (PLA) and acrylo nitrile butadiene styrene (ABS). Polyethylene terephthalate (PET), graphene, polycarbonate (PC) and nylon have wider applicability (Mannoor et al., 2013).

Process parameters have a significant impact on output quality characteristics; hence many researchers have worked on optimizing process parameters by applying conventional and modern techniques (see next section).

Taguchi's quality loss function (QLF) is a reliable and tested technique to ascertain quality deviation from target. Many researchers have applied this method to assess quality loss. Response surface methodology (RSM) is an optimization technique that is widely applied for the purpose (Hansda et al., 2014).

In this research, QLF was combined with RSM to find the quality loss and parametric optimization of an FDM process. The combination of these two

DOI: 10.1201/9781003181644-3

reliable and widely applied techniques will prove to be a powerful, highly appropriate and accurate method of achieving the targets.

3.2 RESEARCH BACKGROUND

Many products printed through FDM have poor strength, surface finish and dimensional accuracy (Sun et al., 2013).The characteristics of products produced by an FDM machine are controlled by the process input parameters, such as raster width, raster angle, the air gap between adjacent tracks and thickness of deposited layers. The common parameters studied by researchers were infill ratio, layer thickness, printing axis, delay time and deposition angle (Dundar et al., 2013). Luzanin et al. (2014) studied the effect of infill ratio, layer thickness and deposition angle on maximum flexural force. Their conclusion was that flexural force was mainly affected by layer thickness and the relation between deposition angle and infill was significant. Purnama et al. (2019) investigated the highest tensile strength and the lowest dimension error, based on optimum parameters (layer thickness, temperatures and raster angles). Their study demonstrated that the tensile strength of printed PLA material was affected by layer thickness, while dimension error was caused by raster angle. Habib et al. (2016) in their experimental investigation found that parameters such as build style, raster width and raster angle have a marked effect on tensile strength. They concluded that the maximum tensile strength was obtained when build style was solid normal, raster width was 0.6064 mm and raster angle was 45°. They also observed that tensile strength depends on build style. Rayegani and Onwubolu (2014), in their study, explored the effect of process parameters on tensile strength. Group method of data handling (GMDH) and differential evolution (DE) techniques were used to find and optimize the process parameter values. Parameters considered were raster angle, raster width, build orientation and air gap, each at two levels. The optimal solution was maximized tensile strength at build orientation 0°; raster angle 50°; raster width 0.2034 mm; and air gap (–) 0.0025 mm. Guerra et al. (2018) explored the contribution of chemical treatment on compressive strength and mechanical behaviour. Their study investigated the effects of input parameters, raster angle, raster width and immersion time on compressive strength. They deployed analysis of variance (ANOVA) to evaluate the effect. Their conclusion was that immersion time around 300 seconds would reduce roughness by 90%, producing good mechanical property parts. Equbal et al. (2018) optimized mechanical properties by developing a mathematical model considering the variables build orientation, layer thickness, raster angle and raster width. ANOVA techniques were employed to find the effect of process parameters. In addition, their research found that improved part strength was the result of increased layer thickness. Wang et al. (2018) found that tensile strength was significantly higher when parts were made in the *z*-axis direction. Also they observed that the weakest strength was observed when printed in the direction perpendicular to the layer. Other researchers (Ang et al., 2016) concluded that mechanical properties and porosity were highly influenced by input environment, such as build orientation, build lay pattern and build layer air gap and raster. Ahn et al. (2003) investigated the effects of process parameters on the tensile and compressive strength of the ABS parts processed by FDM. Parameters considered were model temperature air gap, material colour road width and build orientation. They concluded that process parameters were optimized, and compressive strength and tensile strength were in the ranges of 65–72% and 80–90%.

All research is limited to finding the effects of parameters on mechanical properties and optimizing parameters on different types of strengths. The aim of this study was to find the quality loss as well as parametric optimization for better yields.

3.3 RESEARCH METHODS

3.3.1 Quality Loss Function

Dr Genichi Taguchi in the early 1950s revolutionized the concept of quality, overturning its traditional definition. According to Taguchi, the quality of a product or process is its deviation from its target value of performance (Taguchi & and Jugulum, 2002).

Taguchi developed the term 'loss', defining it as follows: 'quality of a manufactured product is total loss generated by that product to society since the time it is transported'. 'Loss' according to him was the harm caused by the variability of the function and caused by ill effects. An acceptable level of quality has the lowest loss, particularly zero (Taguchi & Jugulum, 2002). The main advantage of QLF lies in its comprehensiveness and simplicity.

The objective function is, thus, designed to keep the mean value closer to the target value and process standard deviation should be kept as low as possible, given by:

$$L(y) = k(y - t)^2 \tag{1}$$

where: L = loss function, y = quality characteristic of the product, t = target value for parameter x and$k$ = quality loss coefficient, expressed as:

$$k = \left(\frac{2}{USL - LSL} \right)^2 \tag{2}$$

Substituting the value of k in equation 1,

$$L(y) = 4\left(\frac{y(x) - t}{USL - LSL} \right)^2 \tag{3}$$

where USL and LSL are the upper and lower specification limits of the quality characteristic respectively.

Loss function is dimensionless quantity and losses due to various characteristics can be added together to find the total loss (TL):

$$TL\left(Y_{1,}Y_{2,\ldots}Y_{n,}X,\ T\right) \ = \ 4\sum_{i=1}^{n}\left(\frac{Yi(X)-t}{USLi-LSLi}\right)^{2} \tag{4}$$

where $Y_{1,}\ Y_{2,\ \ldots}\ Y_{n}$ are quality characteristics (Artiles-León,1996; Roy, 2010; Taguchi & Jugulum, 2002).

3.3.2 Response Surface Methodology

RSM is an analytical technique used to infer results by developing models based on statistical and mathematical concepts. The objective of the analysis is to find the influence of multiple variables on the response and to optimize the variables by performing reverse engineering.

The objective is to maximize the yield (U, V, W) of an FDM process having inputs:

A = infill ratio
B = layer thickness
C = print speed

$$\text{So } U = f(A,B,C)+\varepsilon,\ V = f(A,B,C)+\varepsilon,\ W = f(A,B,C)+\varepsilon \tag{5}$$

where ε = noise or error observed in the response U.

If the expected response can be represented by:

$$E\ (U) = f\ (A,\ B,\ C \ldots\ N) \tag{6}$$

$$E\ (V) = f\ (A,\ B,\ C \ldots\ N) \tag{7}$$

$$E\ (W) = f\ (A,\ B,\ C \ldots\ N) \tag{8}$$

Then their surface can be represented by:

$$N = f\ (A,\ B,\ C) \text{ and is called response} \tag{9}$$

This response surface is represented graphically by plotting:

The N Vs levels of (A, B, C)

The contour or the response is plotted by application of many types of software.

While contouring the plots, lines of constant responses are replicated on the x, y and z axes.

The following steps are needed in development of RSM:

1. To find a suitable approximation of the true functional relationship between Y and the set of independent variables.
2. The approximating function of the first-order model is represented by:

$$Y = \beta_0+ \beta_1 x_1+ \beta_2 x_2+ _{\ldots}\ \beta_n x_n + \varepsilon \tag{10}$$

The RSM is a sequential procedure: in the first stage a domain is found and accurate positioning is done by the second stage.

$$Y = \beta_0 + \beta_1 x_1 + \beta_2 x_2 + \beta_{11} x_1^2 + \beta_{22} x_2^2 + \beta_{12} x_1 x_2 + \ldots \beta_{ij} x_n^n + \varepsilon \tag{11}$$

RSM is applied to determine operating conditions for the system or a region of the factor space in which operating requirements are satisfied effectively (Belwal et al., 2016; Eslami et al., 2016; Šumić et al., 2016).

3.4 RESEARCH GAP

A review of the literature explains a gap between the requirement and availability of research material in the application of QLF in FDM. Although specialists are working every day on it, the efforts require inventiveness, ingenuity and a pioneering spirit that could explore the complex problems of FDM printing technologies, together with novel and advanced techniques that are used as printable substrates. The synergistic combination of materials and technology with the application provides limitless opportunities for the development of new flawless technologies.

Applying QLF would certainly fill the gap between the available and desired results.

3.5 METHODOLOGY

This study was performed in accordance with the objectives of this research. The strategy was to keep the whole methodology simple, straightforward and meaningful. A summary of the methodology is given below.

3.5.1 Experimentation Details

The objective of this research was: (1) to find the loss function; and (2) to optimize the process parameters of an FDM process. The experimentation was designed considering two levels of three parameters; according to study and experience, these three parameters are important and their most popular two levels were studied for their effect on three most sought properties according to the material available. The experiment was designed using full factorial design of the order of $2^3 = 8$. Another four experiments at lower level were performed to confirm the predicted results.

3.5.2 Design of Test Specimen

The most sought mechanical properties are under scanner, hence material testing apex body standards were chosen for printing by FDM. American Standards of Testing Materials (ASTM) D638-10 (type IV) standard test method for tensile properties of plastics was considered, since material used to print in FDM is PLA, a common plastic, widely used in FDM parts (Figure 3.1).

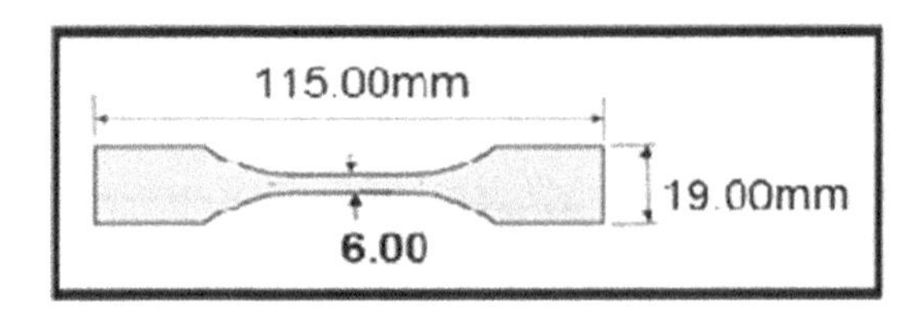

FIGURE 3.1 Design of test specimen.

3.5.3 Process Details

Machine		Material	
Printer	V-SIGMA	Filament	Poly lactic acid
Make	Vertonics Innovations, India	Model	FIBREEL
Volume	390 × 360 × 490 mm	Make	REVER Industries, India
Print speed/ temperature	25–150 mm/180–260°C	Diameter	0.4 mm
Filament	Poly lactic acid, acrylo nitrile butadiene styrene and nylon	Colour	Blue
Nozzle diameter	0.4 mm (F – 1.75 mm)	Temperature	180°C
Layer height	50–300 μm	(Figure 3.2) Bed temperature	60°C

3.5.4 Testing

The specimens were printed in accordance with the printing scheme. Results are recorded and presented in Table 3.1, and the design of experiments and the machines used are shown in Table 3.2.

3.5.5 DATA PROCESSING

3.5.5.1 Loss Function

The loss function is calculated individually and combined by the functions given in the section 3.1 as per equations (3) and (4). Results are detailed in Table 3.3, taking the values of:

$Y_1, Y_{2,\ldots}$ as U, V and W; t = target (mean value of parameters)
$USL_i = 22.58986$ and $LSL_i = 19.37019$ (for ultimate tensile strength: UTS)
$USL_j = 36.32325$ and $LSL_j = 34.09091$ (for compressive strength: CS)
$USL_k = 33.16$ and $LSL_k = 31.70$ (for flexural strength: FS)

TABLE 3.1
Process parameters, their levels and corresponding results

	Process parameters			Performance characteristics		
S. no	Infill ratio (%) (A)	Layer thickness (mm) (B)	Print speed (mm/s) (C)	Ultimate tensile strength (MPa) (U)	Compressive strength (MPa) (V)	Flexural strength (MPa) (W)
1	40	0.1	100	19.37029	34.09091	31.70
2	60	0.1	100	19.39218	36.02716	32.84
3	40	0.1	120	19.38641	36.20184	33.16
4	60	0.1	120	19.38022	35.06032	32.00
5	40	0.2	100	19.38800	34.80881	31.90
6	60	0.2	100	21.70388	36.88042	32.68
7	40	0.2	120	20.37029	36.38044	32.86
8	60	0.2	120	22.58986	38.72600	32.42
9	40	0.1	100	20.82066	36.32325	33.38
10	40	0.1	100	20.00842	36.00233	32.98
11	40	0.1	100	19.84662	35.20032	32.50
12	40	0.1	100	20.50102	35.82232	32.80

TABLE 3.2
Design of testing machines used in the experiments

Testing machine 1		Testing machine 2	
Machine	Universal testing machine	Machine	Hardness tester
Model	AMT, 5A-5B	Model	TRSN-D
Make	Harriss & Tarriss	Make	Fine Manf. Ind.
Maximum load	1000 (kN)	Maximum Load	150(kgf)
Load range	20–1000 (kN)	Load Range	60, 100 & 150(kgf)
Accuracy	±1%	Accuracy	±1%
Initial load range	20 (kN)	Initial load range	10(kgf)
Test performed	Tensile, compressive	Test performed	Rockwell & Brinell
ASTM standards	E83, ISO 9513 (Figures 3.3–3.6)	ISO standards	BS 3846, EN 1002–4

3.5.5.2 Response Surface Methodology

The response surface methodology is invoked by making the calculations in line with section 3.2, equations 5 and 6, taking their respective normal values. To find RSM, the data obtained after testing was evaluated following the procedure explained in section 3.2 in accordance with equations 10 and 11.

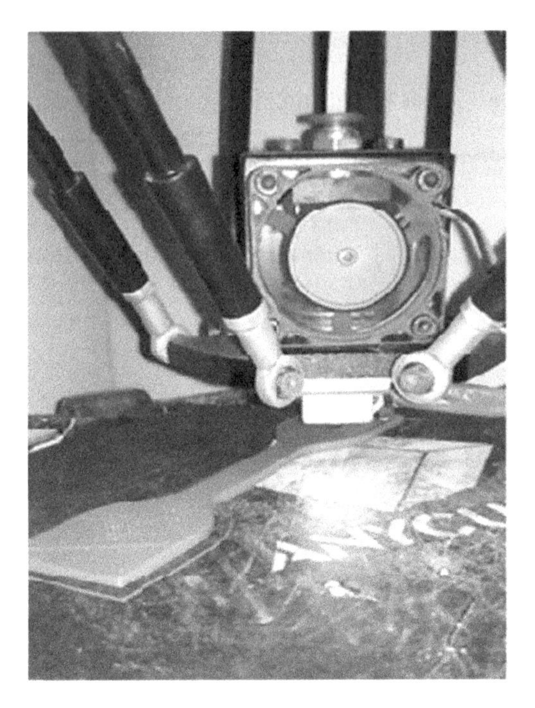

FIGURE 3.2 Work in progress (WIP).

FIGURE 3.3 Universal testing machine (UTM) display.

The following RSM equations have been developed to show the main first-order, second-order and interaction effects of the three process parameters on the three main properties of the FDM process:

$$Y(UTS) = -0.3374 + 0.0055X_1 - 0.0496\,X_2 + 0.0755\,X_3 - 0.00001449\,X_1^2 + 0.2275\,X_2^2 - 0.00551\,X_3^2 + 0.00000250\,X_1X_2 + 0.00049\,X_2X_3 + 0.00005875\,X_1X_3 \quad (12)$$

$$Y(CS) = 8.3374 + -0.2992X_1 - 7.0496\,X_2 - 0.1453\,X_3 - 0.03420\,X_1^2 + 7.5242\,X_2^2 - 0.05743\,X_3^2 + 0.01125\,X_1X_2 + 0.0812\,X_2X_3 + 0.00035\,X_1X_3 \quad (13)$$

$$Y(FS) = -0.3374 + 0.0055X_1 - 0.0496\,X_2 + 0.0755\,X_3 - 0.00001449\,X_1^2 + 0.2275\,X_2^2 - 0.00551\,X_3^2 + 0.00000250\,X_1X_2 + 0.00049\,X_2X_3 + 0.00005875\,X_1X_3 \quad (14)$$

The results are discussed in the next section.

FIGURE 3.4 Universal testing machine display.

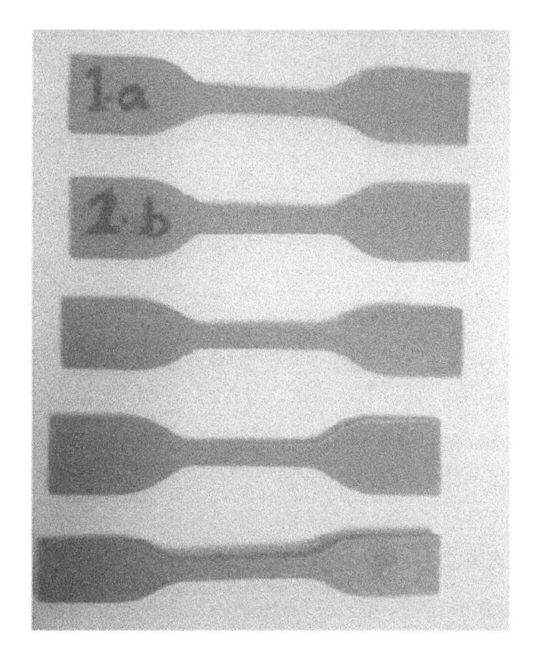

FIGURE 3.5 Printed parts.

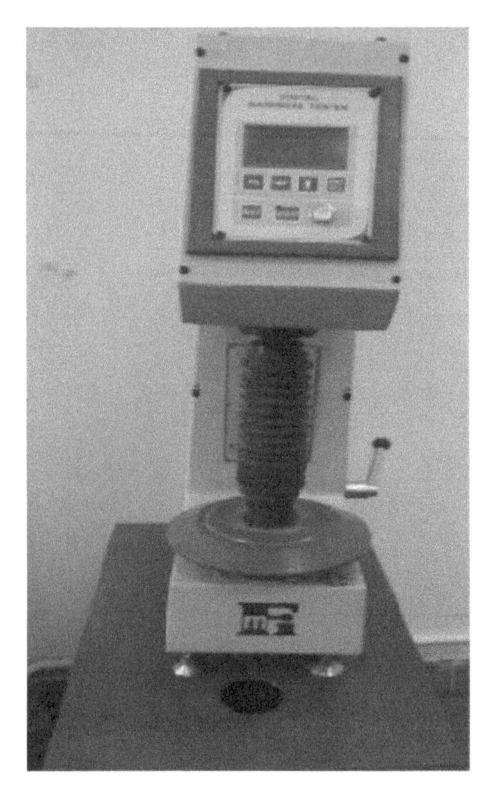

FIGURE 3.6 Hardness tester.

TABLE 3.3
Weight age of individual parameter on mechanical properties and loss

	UTS	CS	FS		Mean	Variance	Expected loss (*Xi*)	*F*(*Xi*)
For IR	0.0355	0.1357	0.1530	For IR	0.323	0.0764	0.1830	0.18
For LT	0.0284	0.1673	0.1673	For LT	0.090	0.2599	0.2598	0.21
For PS	0.0428	0.1972	0.1387	For PS	0.0891	0.7892	0.8813	0.22

UTS, ultimate tensile strength; CS, compressive strength; FS, flexural strength; IR, infill ratio; LT, layer thickness; PS, print speed.

3.6 RESULTS AND DISCUSSION

The results achieved after testing and data were statistically analysed using QLF to find the deviation of the process from its target value. Further analysis was done by applying RSM to identify the effect of process parameters on the mechanical performance of the FDM printed parts. The QLF analysis was based on three factors: (1) UTS; (2) CS; and (3) FS.

The QLF has delivered the three values 0.18, 0.21 and 0.22. This means that the target value of UTS has deviated by 18%, CS value is missed by 21% and FS is also missed by 22%. Together the loss in quality of the process has 20.5% deviation.

Application of QLF and RSM to the test data and subsequent analysis yielded the following results.

1. **Ultimate Tensile Strength**
   (a) **Infill Ratio:** Two levels of infill ratio, 40% and 60%, were considered for study and it was found that a higher level of infill ratio will increase the UTS; the response is the result of greater occupancy of the material in the volume it contained. In the response, there is a significant increase in UTS when going from 40% to 60%.
   (b) **Layer Thickness:** Two levels of layer thickness, 0.1 mm and 0.2 mm, are considered in this study. Again increased layer thickness will directly increase UTS. The increased strength is due to greater internal strength of material with the thick layers.
   (c) **Print Speed:** Decent speeds of 100 and 120 mm/second were considered for this study; it was found that increased speed reduces all types of strength, including UTS, since two adjacent layers have less time to settle down and make a strong bond.
2. **Compressive Strength:**
   (a) **Infill Ratio:** Infill ratio has a direct impact on CS; a higher packing factor is responsible for a more rigid structure and hence CS. Also it was found that 60% infill ratio yields appropriate CS.
   (b) **Layer Thickness:** Similarly to UTS, the CS is a direct function of layer thickness. Layer thickness will affect the CS more than any other factor. The result is loud and clear: 0.2 mm thickness has a higher score than 0.1 mm thickness, considered in this study.
   (c) **Print Speed:** Print speed has an inverse effect on CS. As stated above, it was found that increased speed reduces all types of strength, and CS is not an exception. The probable reason is that consecutive layers have less time to interact, resulting in a poor bond.
3. **Flexural Strength:**
   (a) **Infill Ratio:** FS increases with the infill ratio: 60% infill ratio has a greater impact on FS compared to 40% and the difference is notable. How infill ratio relates to FS is a matter for further study. Little research material is available to help.
   (b) **Layer Thickness:** A higher level of layer thickness is responsible for higher FS: 0.2-mm layer thickness will yield better results on FS.
   (c) **Print Speed:** As far as print speed is concerned, 100 and 120 mm/second were considered and it was found that greater speed reduces all types of strength, including FS. Two adjacent layers have less time to fuse and solidify together and fail to make a strong bond.

Moreover, the optimal set of parameters for maximum UTS, CS and FS is given by:

Infill ratio: 55%, layer thickness: 0.1 mm, print speed: 110 mm/second.

3.7 CONCLUSION

This study has developed a new method of finding functional loss of the process and hence deviation of the quality along with exploring the optimum level of parameters for nominally the best type of optimization. The deviation in the process parameters from its nominal values contributed to loss of a quality. The evaluation of reliability function in this process of proposed method is very simple and appropriate to the derived results.

This method has some limitations; the loss function approach applied in this study was not applied to the highest level of parameters. Thus, it is necessary to develop a way to apply it to the highest level.

REFERENCES

Ang, K.C., Leong, K.F., Chua, C.K. and Chandrasekaran, M., 2016. Investigation of the mechanical properties and porosity relationships in fabricated porous structures. *Rapid Prototyping Journal*, *12*, 100–105.

Artiles-León, N., 1996. A pragmatic approach to multiple-response problems using loss functions. *Quality Engineering*, *9*(2), 213–220.

Belwal, T., Dhyani, P., Bhatt, I.D., Rawal, R.S. and Pande, V., 2016. Optimization extraction conditions using response surface methodology (RSM). *Food Chemistry*, *207*, 115–124.

Du, X., Fu, S. and Zhu, Y., 2018. 3D printing of ceramic-based scaffolds for bone tissue engineering: an overview. *Journal of Materials Chemistry B*, *6*(27), 4397–4412.

Dundar, M.A. and Ayorinde, E., 2013. Plate vibration and geometry study of impact-control ABS. *ASME 2013 International Mechanical Engineering Congress and Exposition*.

Equbal, A., Equbal, M.A., Sood, A.K., Pranav, R. and Equbal, M.I., 2018. A review and reflection of FDM parts. In *IOP Conference: Materials Science Engineering*, Vol. 455, p. 113.

Eslami, A., Asadi, A., Meserghani, M. and Bahrami, H., 2016. Optimization of sono degradation of amoxicillin by response surface methodology (RSM). *Journal of Molecular Liquids*, *222*, 739–744.

Guerra, M.G., Volpone, C., Galantucci, L.M. and Percoco, G., 2018. Photogrammetric measurements of 3D printed microfluidic devices. *Additive Manufacturing*, *21*, 53–62.

Habib, F.N., Nikzad, M., Masood, S.H. and Saifullah, A.B.M., 2016. Design and development of scaffolds for tissue engineering. *3D Printing and Additive Manufacturing*, *3*(2), 119–127.

Hansda, S. and Banerjee, S., 2014. Multi characteristics optimization using Taguchi quality loss function. In *Proceedings of 5th International Manufacturing Technology, Design and Research Conference*.

Lužanin, O., Movrin, D. and Plančak, M., 2013. Experimental investigation of extrusion speed and temperature effects on arithmetic mean surface roughness in FDM built specimens. *Journal for Technology of Plasticity*, *38*(2), 179–190.

Macdonald, E. et al., 2014. 3D printing for the rapid prototyping of structural electronics. *IEEE Access*, *2*, 234–242.

Mannoor, M.S., et al., 2013. 3D printed bionic ears. *Nano Letters*, *13*(6), 2634–2639.

Purnama, I.L.I., Tontowi, A.E. and Herianto, H., 2019. 3D Image reconstruction with single-slice. *International Biomedical Instrumentation and Technology Conference*, Vol. 1.

Rayegani, F. and Onwubolu, G.C., 2014. Fused deposition modelling (FDM) process parameter prediction and optimization using group method for data handling (GMDH) and differential evolution (DE). *The International Journal of Advanced Manufacturing Technology*, *73*(1–4), 509–519.

Roy, R.K., 2010. A Primer on the Taguchi Method. Society of Manufacturing Engineers.

Šumić, Z., Vakula, A., Tepić, A., Čakarević, J., Vitas, J. and Pavlić, B., 2016. Modeling and optimization by response surface methodology (RSM). *Food Chemistry*, *203*, 465–475.

Sun, K, et al., 2013. 3D printing of interdigitated Li-Ion microbattery architectures. *Advanced Materials, 25*(33), 4539–4543.

Taguchi, G. and Jugulum, R., 2002. The Mahalanobis-Taguchi Strategy: A Pattern Technology System. John Wiley.

Wang, L., Zhang, M., Bhandari, B., and Yang, C., 2018. Investigation on fish surimi gel as promising food material for 3D printing. *Journal of Food Engineering*, *220*, 101–108.

Zheng, Y.X., Yu, D.F., Zhao, J.G., Wu, Y.L. and Zheng, B., 2016. 3D printout models vs. 3D-rendered images. *Journal of Surgical Education*, *73*(3), 518–523.

Zhu, W., Ma, X., Gou, M., Mei, D., Zhang, K. and Chen, S., 2016. 3D printing of functional biomaterials for tissue engineering. *Current Opinion in Biotechnology*, *40*, 103–112.

4 Effect of Physical Attributes of Coconut on Effective Husk Separation

A Review

Abi Varghese, Joby G. David,
Mebin Toms Mathew, Mijo P. Saji, and
Sambhu Nair V.S.

Department of Mechanical Engineering, Amal Jyothi College of Engineering, Koovappally, Kanjirappally, Kottayam, Kerala, India

4.1 INTRODUCTION

4.1.1 BACKGROUND

A fruit with many uses, coconut grows on the tree *Cocos nucifera* belonging to the Arecaceae palm family (Nair, 2010). The hardest part of the nut is the endocarp and it has three germination pores. The fleshy middle layer, the mesocarp, is made of coir fibres. Once the husk is removed, pores are clearly visible. Testa is present in the inside wall of the endocarp which contains the fleshy, white edible part of the seed, a thick albuminous endosperm. During growth the husk and shell become stiffer (DebMandal and Mandal, 2011). The shell is separated from the endosperm by a 1–2-cm thick brown layer of testa. A normal coconut weighs 1.44 kg and by the time a coconut falls naturally, the coir has become drier and softer and the husk has become brown (Thampan, 1981).

There are three parts to the operations in farm mechanization: (1) pre-harvesting; (2) harvesting; and (3) post-harvesting (Varghese and Jacob, 2014). The pre-harvesting operation comprises seed selection, land preparation, sowing, and irrigation. The fruit or yield from a specific plant is harvested in the harvesting operation. For consumption, storage, or preservation and transportation the post-harvesting operation is carried out, comprising husking, breaking, shelling, and drying.

Husking is the most tedious amongst various post-harvest operations. Traditional tools used in husking are chopping knives or machetes, crowbars

DOI: 10.1201/9781003181644-4

(paara), and axes (Varghese and Jacob, 2014). To mechanize this operation many machines have been invented all over the world, but only a few have remained in the market; some disappeared and for many reasons others have not yet reached the market. Mechanization of these processes is done after establishing a relationship between the physical attributes of the coconut and that of the machinery involved. The dimensions of the machine parts, their weight, geometry, and material selection have a direct relation to the coconut's physical attributes.

As already mentioned, all over the world coconut is a popular drupe amongst various others. To mechanize the post-processing operations several machines have been invented and used in various parts of the world. The current study is concentrated on the relationship between the various physical attributes of the coconut and their effect on husking operations.

4.1.2 Research Objectives

- Study of various coconut husking machines, both manual and power-operated
- Study of various physical attributes of the coconut
- Effect of these physical attributes on the design of coconut husking machines.

4.2 METHODS

This review is concentrated on two research questions: (1) What are the various physical attributes of the coconut that need to be considered for effective husk separation? and (2) How do the various physical attributes of the coconut influence the mechanism for effective husking? The various physical attributes under study are size, weight of coconut, shell diameter, moisture content, and husk thickness. These attributes were particularly chosen because of their contribution to the development of coconut husking mechanisms. The existence of a mutual relationship between these attributes is essential for an effective husking mechanism. A mismatch could lead to a significant reduction in the efficiency of the husking process. Variations in the above-mentioned physical attributes of coconut can be seen across the world. The influence of each parameter on successful husking can only be analysed in a study on the existing mechanisms and establishing a relationship between their various physical attributes.

4.2.1 Review Strategy

Data analysis was to be made on the basis of collected information regarding the physical attributes of coconut and the part design of existing machinery for husking. For this, papers involving the topics were taken as a reference. The idea was to establish a clear relationship between the various physical attributes and the design of mechanisms.

The review was done in such a manner that it could be used for future developments in the field of coconut husking machinery and other coconut post-harvesting operations.

4.2.2 Screening

The collected articles were separately screened by two reviewers. The other three reviewers screened the articles at the title, abstract, and full-text pages. During various levels of screening, differences were examined and fixed through consensus.

4.2.3 Data Extraction

Information needed to address the primary research questions was acquired by data extraction. Information collected included: author's name and date of publication; coconut feeding mode; shape of blade involved; orientation of blade; actuating mechanism; husking mechanism; movement of the lever involved in the mechanism; the maximum, minimum, and mean values of various physical attributes of green and dry coconuts, such as: length, diameter, perimeter, weight, husk thickness, and husk separating force.

4.3 RESULTS AND DISCUSSION

4.3.1 Search Results

The search produced 26 articles, 24 related to existing coconut husking mechanisms and two related to the physical attributes of coconuts.

4.3.2 Physical Attributes of Coconut

The review was based on data collected from 35 varieties of coconuts (Varghese et al., 2016). Green as well as dry coconuts were considered and the physical attributes under study include: length, diameter, weight, perimeter, husk thickness at different distances from the pedicel end, and the husk separating force required. The minimum, maximum,and average values of the above-mentioned attributes were noted (Table 4.1).

4.3.3 Analysis of Coconut Husking Mechanisms

The analysis involved manually operated as well as power-operated husking mechanisms. Ten mechanisms in the former category and six mechanisms in the latter category were analysed. The data collected include: name of inventor and year of introduction, coconut feeding mode, shape of blade, orientation of blade, movement of lever and actuating mechanism, and husking mechanism in case of power-operated mechanisms (Tables 4.2 and 4.3).

4.3.4 Implementation of Coconut Physical Attributes Into Husking Mechanisms

It is clear that most of the existing machinery has wedge-shaped blades, whose dimensions depend upon the husk thickness of the coconut to be impaled on it.

TABLE 4.1
Physical attributes of Indian coconut (Varghese et al., 2016)

Physical property		Green coconut			Dry coconut		
		Minimum	Maximum	Mean	Minimum	Maximum	Mean
Length (mm)		230.29	287.65	248.89	220.73	293.44	246.98
Diameter (mm)		138.31	287.42	200	112.24	226.85	179.57
Perimeter (mm)		427.89	687.98	583.06	436.43	687.67	565.93
Weight (kg)		0.75	1.95	1.3	0.75	1.26	0.98
	At apex	1.7	3.3	2.7	2.7	4	3.3
	At pedicel end	3.3	8.1	5.4	4.8	7.5	6.4
Husk thickness (mm)	At ¼ distance from pedicel end	2	5.2	3.3	2	3.6	2.9
	At ½ distance from pedicel end	0.7	3.3	2	1.8	3	2.3
	At ¾ distance from pedicel end	0.8	2.8	2	1.2	3	1.6
Mechanical property							
Husk separating force (kN)		600.3	907.2	745	635	964.2	856.7

TABLE 4.2
Manually operated husking mechanisms

Sl. No.	Tool	Inventor	Year	Coconut feeding mode	Shape of blade	Orientation of blade	Movement of lever
1	Coconut husk-removing tool (Waters, 1949)	Cecil P. Waters	1949	Held on a platform	Thin wedges	Juxtaposed or closed	Downward (hand lever)
2	Coconut husking machine (Titmas & Hickish, 1929)	R. W. Titmas and R. S. Hickish	1929	Hand-held	Vertical straight-sided	Juxtaposed upright	Downward (foot lever)
3	Coconut spanner (Muraleedharan, 1996)	Unknown	1996	Placed on a platform	Sharpened tongue	Juxtaposed or closed	Downward (hand lever)
4	Coconut husk-removing tool (Hill, 1983)	Edward D. Hill	1983	Placed in a bowl in upright position	Twin wedge	Juxtaposed or closed	Downward (hand lever)
5	Mini coconut de-husker (Ganesan & Gothandapani, 1995)	Ganesan and Gothandapani	1995	Kept on platform	Tong-like tool	Juxtaposed or closed	Downward (hand lever)
6	KAU coconut husking machine (Muraleedharan, 1998)	Aboobekkar and Narayanan	1998	Hand-held	Twin wedge	Juxtaposed or closed	Downward (foot lever)
7	Keramithra coconut husking tool (Jacob & Bastain, 1995)	J. Jippu and B. Joby	1998	Hand-held impaling	Twin wedge	Juxtaposed upright	Upward (hand lever)
8	Ce Co Co coconut de-husker (CeCoCo;Jacob, 1995)	CeCoCo, Japan	Unknown	Hand-held impaling	Thin wedges	Juxtaposed upright	Downward (foot lever)
9	Coconut de-husking machine (Dinanath et al., 1987)	Chandra Dinanath, Chaguanas	1987	Roller feeding	Sharpened spikes	Spikes attached to rollers	No lever
10	Apparatus for removing fibre from coconut (Marot, 1911)	Rene Marot	1911	Placed on support	Sharpened knives	Knives arranged in vertical planes	Downward (hand lever)

TABLE 4.3
Power-operated husking mechanisms

Sl. No.	Tool	Inventor	Year	Actuating mechanism	Husking mechanism	Coconut feeding mode
1	Twin blade coconut husking machine (Varghese & Jacob, (2014)	Unknown	Unknown	1.5 hp motor	Rotation of blades	Hand-held
2	Rotary coconut de-husker (Muhammad, 2002)	Kelappaji College of Agricultural Engineering	2002	Motor drive	Shear force exerted by blades of rotating drum and concave	Hand feeding with guide mechanism
3	Hydraulic coconut de-husking machine (Coconut Development Board, 2011)	Fletchers Engineering, Australia	2011	Electric and hydraulic drives	Lifting the holder along with arm	Holding mechanism
4	Mechanical coconut husking machine (Santhi et. al., 2006)	Kelappaji College of Agricultural Engineering	2006	1.5-hp, single-phase AC motor	Curved sharp hook-like knives on powered rollers	Hand-held
5	Twin-blade power-operated coconut-husking machine (Aneesh et. al., 2009)	Kelappaji College of Agricultural Engineering	2009	Cam and follower mechanism	Hinged movable blade and stationary blade	Hand feeding
6	Continuous power-operated coconut husking machine (Anu et al., 2012)	Kelappaji College of Agricultural Engineering	2012	Motor drive	Shear force exerted by blades of rotating drum and concave	Hand feeding

A slight variation in dimensions could lead to improper husking with lower efficiency. The diameter and length of the coconut influence the design of the platform on which the coconut to be impaled is placed or positioned in the case of manually operated mechanisms; and in the case of power-operated mechanisms, these attributes are necessary for the calculation of shear force exerted by blades of rotating drum and concave. These also influence the coconut feeding mode. The magnitude of weight of the coconut influences the force required to be applied on the lever as well as the impaling force needed to pierce the coconut on to the blades of the husking mechanism. The weight also influences the actuating mechanism to be used in the power-operated mechanisms.

4.4 CONCLUSION

Through this review it can be concluded that the physical attributes of coconut perform an important role in the development of coconut husking technologies. Implementation of these attributes into the design not only improves the efficiency of husking but also leads to simplification of design. The establishment of interrelationships between the physical attributes of coconut and the design of the husking machinery is vital for successful operation. It is only through such an establishment that one can reduce the difficulty involved in the design of husking mechanisms and the cost involved through design simplification.

REFERENCES

Aneesh, M., Anu, S. C. and Shabeena, P. K., 2009. Development of a Power Operated Coconut Husking Machine. B. Tech project report submitted to the Kelappaji College of Agricultural Engineering and Technology (KCAET), Tavanur.

Anu, S. C., 2012. Development and Testing of a Continuous Power-Operated Coconut Husker. M. Tech project report submitted to the Kelappaji College of Agricultural Engineering and Technology (KCAET), Tavanur.

CeCoCo. Pamphlet on CeCoCo OKIMI Coconut Cracker (De-husker), Japan.

Coconut Development Board, 2011. Hydraulic Coconut De-husking Machine. Coconut Development Board/De-husking Machine.

DebMandal, M. and Mandal, S., 2011. Coconut (*Cocos nucifera* L.: Arecaceae): in health promotion and disease prevention. Asian Pacific Journal of Tropical Medicine, 4: 241–247.

Dinanath, C. 1987. Coconut De-husking Machine. US patent no. 4,708,056.

Ganesan, V. and Gothandapani, L., 1995. Mini coconut dehusker. TNAU News Letter, 25(7): 3.

Hill, E. D., 1983. Coconut Husk Removing Tool. US patent no. 4,383,479.

Jacob, J., 1999. Some manually operated coconut husking tools: a comparison. Indian Coconut Journal, 5: 56–59.

Jacob, J. and Basitian, J., 1995. Coconut Husking Tool. Indian patent no. 192670.

Marot, R., 1911. Apparatus for Removing Fiber from Coconut. US patent no. 983631.

Muhammad, C. P., 2002. Final Report of the Project on 'Development of Equipment and Technology for Pre-Processing of Coconut'. Kelappaji College of Agricultural Engineering and Technology, Tavanur, India.

Muraleedharan, T., 1996. Coconut spanner for husking. The Matrubhoomi Daily, 23 January, Palakkad edition, Palakkad, Kerala.

Muraleedharan, T., 1998. A new tool for husking. The Malayala Manorama Daily, 18 November, Trichur edition, Trichur, Kerala.

Nair, K., 2010. The coconut palm (*Cocos nucifera* L.). In: The Agronomy and Economy of Important Tree Crops of the Developing World, pp. 67–109. https://doi.org/10.1016/B978-0-12-384677-8.00003-5

Santhi, M. M., Sudheer, K. P. and Prince, M. V., 2006. Evaluation of Power Operated Coconut De-Husking Machine. Annual Report 2005–06, All India Coordinated Research Project on Post-Harvest Technology (ICAR), Department of Post-Harvest Technology and Agricultural Processing, KCAET for presentation in the annual report held at Bhubaneswar, Orissa, pp. 6–13.

Thampan, P. K., 1981. Handbook on Coconut Palm. New Delhi: IBH Publishing.

Titmas, R. W. and Hicklish, R. S., 1929. Coconut Husking Machine. US patent no. 1724732.

Varghese, A. and Jacob, J. 2014. A review on coconut husking machines. Proceedings of the International Conference on Emerging Trends in Engineering and Management (ICETEM 14), vol. 5(3), pp. 68–78.

Varghese, A., Francis, K. and Jacob, J., 2016. A study of physical and mechanical properties of the Indian coconut for efficient de-husking. Journal of Natural Fibers 14(3): 390–399. doi:10.1080/15440478.2016.1212760.

Waters, C. P., 1949. Coconut Husk Removal Tool. US patent no. 674,305.

5 Selection of Features and Classifier for Controlling Prosthetic Devices

Keerti and Karan Veer

Department of Instrumentation and Control Engineering, Dr. Bhimrao Ambedkar National Institute of Technology, Jalandhar, India

5.1 INTRODUCTION

A few decades ago, the area of electromyogram (EMG) signal research field was limited. At that time research was mainly focused on forearm gestures controlled using the muscles of the upper arm. Nowadays due to the new advanced technology, single figure movements can be controlled using the myoelectrical controlling devices. Currently a single device can control different degrees of freedom (DOF). This advanced control is possible by combining digital devices with myoelectrical signal. To use other types of hand movement in EMG-dependent control, investigators have examined the isolation and reproduction of different hand movements. In the present era of research, results show that not only finger, but also the joints, can be successfully identified from each other utilizing the surface EMG (sEMG) signal [1–3].

In recent years, sEMG signal has become an attractive field for researchers using human–machine interface (HMI) and rehabilitation-type applications. Peripheral equipment related to hand gesture activity can be controlled by myoelectrical signal. Hand movement can be detected by sEMG, using the sensing device to collect the myoelectrical signal from different muscles [4, 5].

With regard to precise identification of user intention based on sEMG signal, there is a major problem in obtaining measurement of the sEMG signal for myoelectrical control. Since the 1970s investigators have analysed different hand gestures, including the flexibility extension and wrist extension, by acquiring hand signals [6, 7]. From current research work we understand that the pattern recognition (PR) rate is above 85% but there are still difficulties that can be minimized using current techniques.

Implementation of a myoelectrical system with more than 14 sensors is not feasible, because as the number of sensors increases, the complexity of

DOI: 10.1201/9781003181644-5

signal acquisition system also increases, creating difficulties for the subject in performing different hand gesture. Also, a larger number of sensors increases the complexity in understanding the sEMG signal which is related to different hand movements. In addition, problems arise when attaching the sensing devices to the user's hand while doing the movements. Future work is needed on proper recognition of signal action in signal with some sensors to achieve a multi-sensing DOF bioelectrical control signal.

The sEMG signal is a myoelectrical signal which is collected from the hand muscles. This biomedical signal is used for industries such as gaming, aeronautical and biomedical. The electrical potential generated by the muscles is used as the input signal for clinical purposes in biomedical industries. In this literature we use techniques for the analysis of EMG signal in biomedical and robotics industries. EMG potential value lies in the 0–10-mV, 20–500-Hz frequency range. The term 'electromyography' is derived from three words: electro, meaning electricity; myo, meaning muscles; and grapho, meaning writing. sEMG signal can be collected by invasive and non-invasive electrodes.

In this chapter we apply the different techniques for PR of different hand gestures and movements. We performed the experiment with six different hand gestures. This chapter also consists time domain and power spectrum analysis for each set of hand gestures.

5.2 METHODOLOGY

5.2.1 Data acquisition

Raw EMG signal waacquired from six hand gestures. Each movement was performed for 1 minute and a 30-second time gap was maintained between hand motions. Carpi radialis longus muscles with the gel are used to reduce skin-to-electrode impedance when recording the sEMG signal. The sampling rate was 1 kHZ and gain was 1000 times, so, 1000 samples were recorded in each second. The six sets of hand movements were wrist and figure movements.

For this research the six hand gestures were forearm supination (FS), hand closed (HC), cylindrical grip (CG), wrist left (WL), wrist right (WR) and peace sign (PS) (Figure 5.1). While performing the hand gesture the subject was seated on a chair and placed his/her hand on a flat table [8].

5.2.1.1 Electrodes

There are two types of electrode: invasive and non-invasive. Non-invasive electrodes are also known as surface electrodes, whereas invasive electrodes are known as intramuscular electrodes. Muscle potential is converted to electronic potential by electrodes. In this study we used surface or non-invasive electrodes. The electrode was located on the subject's right-hand flexor carpi radial muscle. The raw EMG signal for all six sets of hand movements is shown in Figure 5.2. The electrode was placed on the skin of the subject's hand, to collect

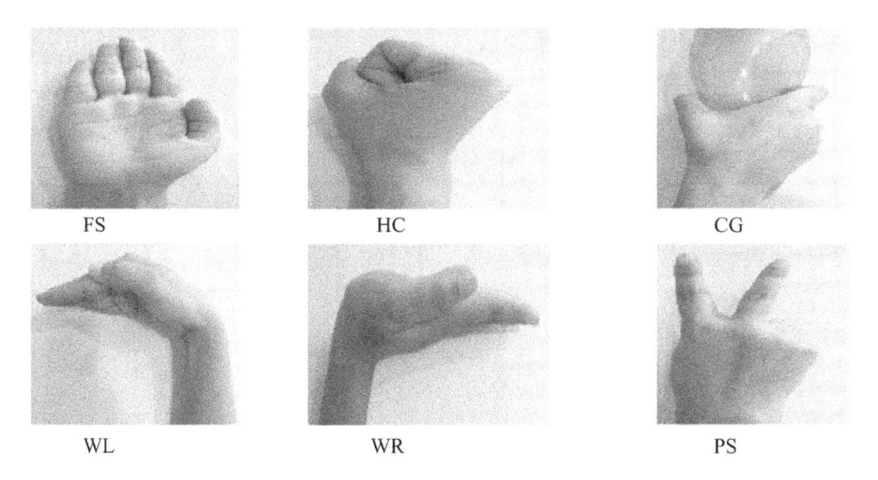

FIGURE 5.1 The six different hand gestures. FS, forearm supination; HC, hand close; CG, cylindrical grip; WL, wrist left; WR, wrist right; PS, peace sign.

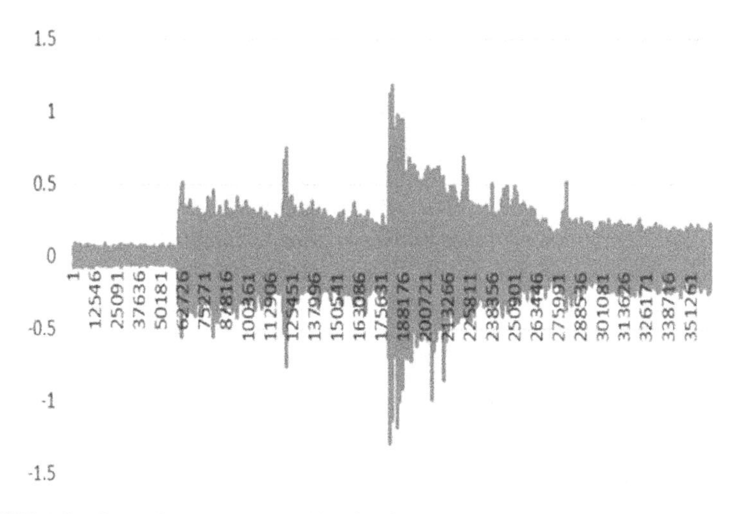

FIGURE 5.2 Raw electromyogram (EMG) signal.

the myoelectrical signal. Gel was applied before the electrode was attached to the hand as the gel reduces skin-to-electrode impedance [9]. Bioelectrical potential was collected from the surface. The electrode placement site was properly cleaned before electrode attachment. For data collection we used disposable electrodes, made predominantly of Ag/AgCl. A When needle electrodes are used, experienced supervision is needed and needle electrodes arestly inserted into a small section of skin. So, data collection using surface electrodes is more demanding compared to invasive electrode. The size of electrode did not have a great influence on sEMG signal [10].

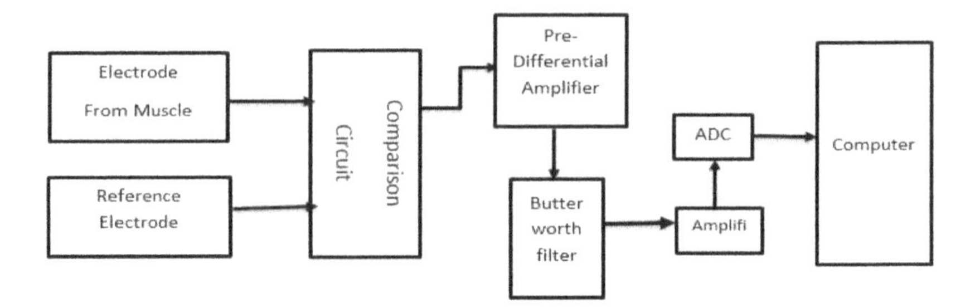

FIGURE 5.3 Data acquisition system for electromyogram (EMG) signal. ADC, analogue-to-digital converter.

5.2.1.2 Rectifier and Filter

The differences between two electrode voltages were taken after recording the signal from the electrodes (Figure 5.3). The entire process of myoelectrical signal acquisition is shown in this figure. After the muscle's ion potential collection, the whole signal is passed through the differential amplifier and the magnitude of the myoelectrical signal is very low. sEMG signal has an amplitude range 0–10 mV. So, it requires pre-amplification. The EMG signal also collects noise while the signal is passing through muscle tissue. Many artifacts are combined with the raw sEMG signal. To filter the unwanted artifacts, we used the Butterworth filter with a sampling frequency of 500 Hz; 20 Hz was selected as the lower frequency. For the EMG signal we mainly used and analogue filter such as bandpass filter or low-pass filter. Filtration is also required to remove unwanted noise signal such as electromagnetic interference signal. These signals combine with the actual EMG signal, making it hard to recognize the actual signal [11]. The EMG signal was rectified and then passed through a digital high-pass or low-pass filter to remove movement and other artifacts. Signal filtration improves the quality of the signal and increases the signal-to-noise ratio [12].

5.2.2 Feature Extraction

To reduce data dimensionality, we found out the features of the sEMG signal; it also helps in the recognition of different sets of hand movements.

sEMG signal contains information such as control and morphological information. The magnitude and shape parameter of the waveform aremorphological information (MI) and action potential and firing control are related to the control information of the sEMG signal. To evaluate the potential variation in magnitude, features can be extracted from the myoelectrical signal, such as time domain, frequency domain and time–frequency domain. The myoelectrical signal is a non-stationary signal. So, the accuracy of the results depends on the selection of the feature of EMG signal. Feature output work, such as the input of testing and training of the classifiers and good selection of features, can improve the performance of a classifier [13, 14]. Before feature extraction normalization is applied to increase the performance of the classifier [15, 16].

Earlier papers have described the different signalling methods, such as autoregression method, wavelet transform (WT) and discrete Fourier transform method. All of these methods can be used with the sEMG signal [17–19]. In this research we used time domain features such as root mean square (RMS) and mean absolute value (MAV).

(a) Root Mean Square Value

The RMS is a statistical calculation of the vastness of different quantities. When the magnitude of the sEMG signal has both negative and positive portions then it is preferable to use this feature.

A higher sEMG signal has a higher RMS value (Figure 5.4). The RMS value of the HC gesture is highest among the all hand movements, whereas the PS hand gesture has the lowest value (Table 5.1).

$$\text{RMS} = \sqrt{\left(\frac{x_1^2 + x_2^2 + x_3^2 \ldots\ldots\ldots\ldots\ldots + x_n^2}{n}\right)} \tag{1}$$

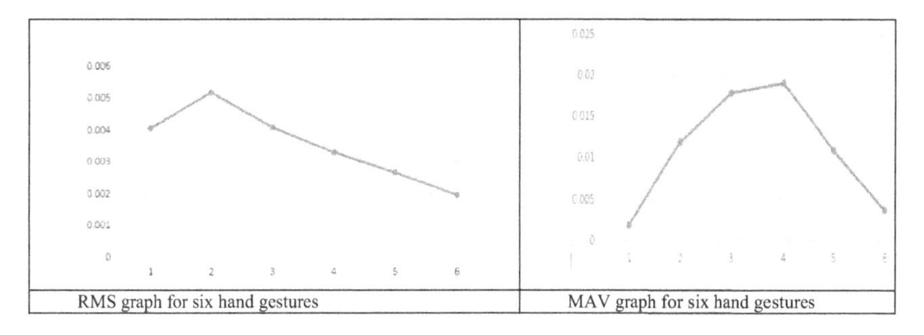

FIGURE 5.4 Root mean square (RMS) and mean absolute value (MAV) graph for hand gestures.

TABLE 5.1
Root mean square (RMS) and mean absolute value (MAV) for six hand gestures

Subject 1	RMS	MAV
1st movement	0.00407	0.001873
2nd movement	0.005197	0.012032
3rd movement	0.004087	0.017952
4th movement	0.003322	0.019093
5th movement	0.002666	0.010936
6th movement	0.00199	0.003788

(b) Mean Absolute Value

MAV is used to extract the average absolute value of sEMG signal in the form of fragments. From Figure 5.4 it can be seen that WL hand gesture has the highest and FC hand gesture has the lowest MAV. Average rectified value (ARV) is also similar to the MAV. After rectification, the area under the EMG signal is shown, meaning that it converts all negative values into positive values.

$$\mathrm{MAV} = \frac{1}{N}\sum_{n=1}^{N} x(n) \tag{2}$$

(c) Power Spectrum Density

The power spectrum density defines the distribution of power into the component of frequency which consists of the EMG signal, using the power spectrum of time series. If a signal has a finite amount of energy then the energy spectral density can be computed. In this study the Hamming window was used to estimate power spectral density (PSD). PSD techniques divide the sEMG signal into a fixed number of time periods. From Figure 5.5 the difference can be estimated between all PSD graphs for each hand movement. The area under the CG movement hand graph is greater in comparison to other PSD; this means CG movement has more power under the curve.

The power spectrum for all six hand movements is shown in Figure 5.5. PSD detects the periodicity of EMG signal and gives the highest peak at the fundamental frequency. It also contains parametric and non-parametric methods, which methods differentiate the real signal's epoch length.

5.3 RESULTS

5.3.1 Result of ANOVA Techniques

ANOVA means analysis of variance. There are many types of ANOVA technique, such as wavelet-based ANOVA. Wavelet-based ANOVA is mainly used for neurophysiological signals.

ANOVA techniques are used to find out the statistically significant difference between the mean of three more number of unrelated groups. It compares the mean within the group and between groups, as well as finding out the significant difference. It is mainly used to test the null hypothesis (NH). It does not compare multiple tests. In this study the authors used ANOVA to compare the dependent variable which is normally distributed. The independent variable is decided by the type of ANOVA but the dependent variable mostly remains one. For example, one-way ANOVA has two independent variables or factors and one dependent variable, whereas two-way ANOVA has two independent variables and one dependent variable. Differences between the group and within the group are known as variability. In this study six independent variables were used. By

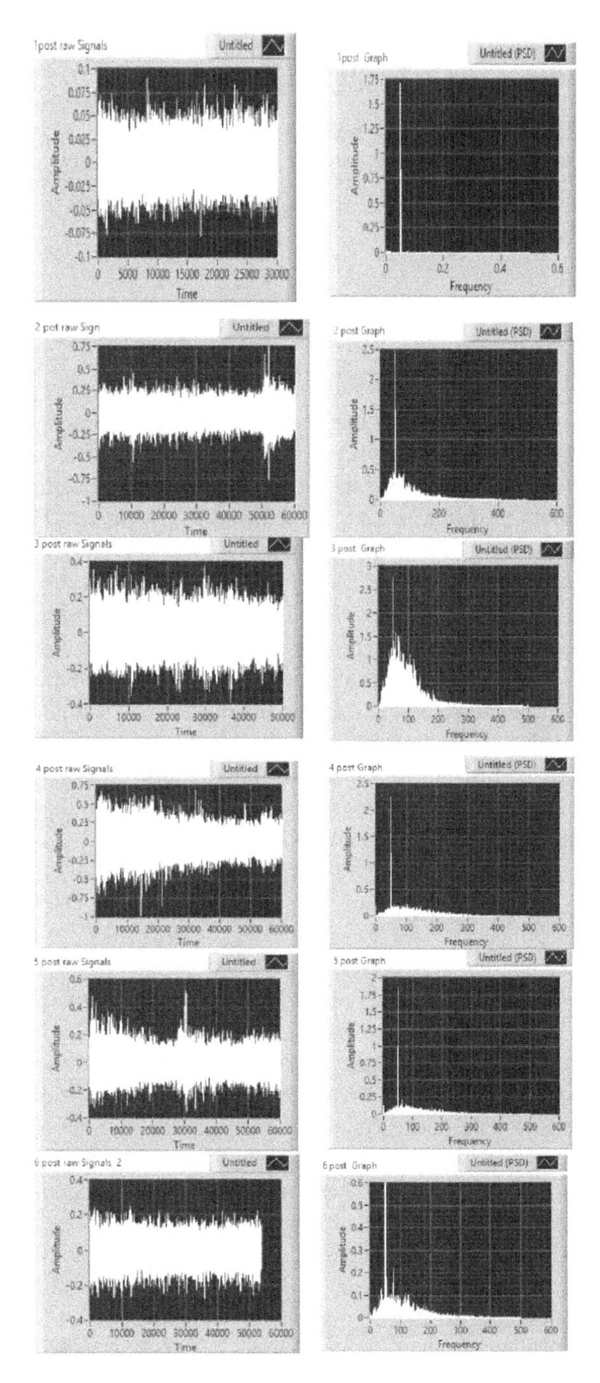

FIGURE 5.5 Power spectral density (PSD) and raw electromyogram (EMG) signal graphs for all six hand movements.

TABLE 5.2
Analysis of variance results for six hand gestures

Source of variation	Sum of square	Degrees of freedom	Mean square	Fisher ratio	Significance (p)	Critical value (F)
SSB	0 .00407	11	0 .364	2.752	0.007	0.662
SSW	0.00187	53	0.0018			
Total sum of square	0.00594	64				

SSB, sum of squares between groups; SSW, sum of squares within groups.

comparing the mean of all the six hand gesture their effectiveness was evaluated. Two estimations, sum of square within group (SSW) and sum of square between groups (SSB) were made. The F-ratio is used to find out the difference between the mean, whether this difference is significant or just fluctuates. The ata are more significant if the F-ratio value is higher [20]. Since the estimate of SSW is smaller than the SSB (0.00407 > 0.00187), it can be concluded that the test statistic is significant. Mean square error should be as small as possible for greater accuracy of ANOVA. Here the mean square error is 0. 001. So, it can be concluded that the given data are significant and all fluctuation of data has been removed.

From Table 5.2 it can be concluded that the Fisher ratio is higher than the critical value. The value of $p < 0.05$.

So, the means of SSB and SSW are different and this ANOVA technique is able to differentiate between the magnitude and shape of all the data relating to the different hand movements of the EMG signal.

5.3.2 Result of Classifier

The K-nearest neighbour (KNN) classifier is a non-parametric PR classifier which is widely use at present. The KNN classifier is researched by Cover and Hart. It is widely used because of it is less complex but highly effective. So, it is used in classification problems. Where there is less information or knowledge about the data then this is the best choice of classification method. The class assigned to a pattern is the class of the closest pattern known to the system, measured based on the distance defined by the feature space. Inthis space, each pattern defines an area (the Voronoi area). When distances are classical Euclidean distances, this is the Voronoi region limited by a linear border.

Currently, it is the most widely used classification algorithm. In this algorithm the testing data were compared with the training data and the pattern was recorded, as well as the K-training record. This algorithm is closely defined by the distance metric term which is also known as Euclidean distance. Six hand motions were defined by the six classes and fivefold cross-validation was used for PR of the EMG signal. The receiver operating characteristic (ROC) curve is shown in Figure 5.6 figure, as well as the area under the curve (AUC). The

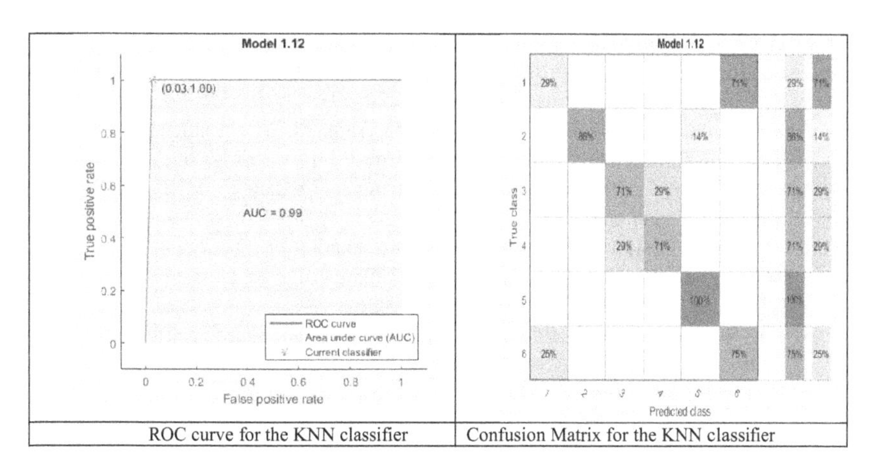

FIGURE 5.6 Receiver operating characteristic (ROC) curve and confusion matrix for the K-nearest neighbor (KNN) classifier. AUC, area under the curve.

true- and false-positive rate of the trained classifier can be seen. In Figure 5.6 there is a value of 0.99, indicating, that the 99% observation is correct. This ROC curve is drawn between the second and fourth hand motion. The AUC shows the quality of the classifier: the greater the AUC, the more accurate the classifier. Accuracy of individual classes is shown by the confusion matrix. In the confusion matrix predicate classes are shown by columns and true classes is shown by rows. PR for the fifth class is higher and for the first class is lower compared to other classes. The overall accuracy of the KNN classifier for PR is 87% with a training time of 0.3838 seconds.

5.4 CONCLUSION

In this study, for the PR of different hand gestures we used the KNN classifier and the outcome of RMS and MAV for the input of the KNN classifier, resulting in an overall accuracy of the classifier of 87%. This technique provides a uniformly better performance. The recorded sEMG signals from the hand muscles for different hand motions were used to control a robot hand and prosthetic device with various DOF. The PR result from the KNN classifier can easily be implemented in various medical applications. In this chapter the time domain feature and ANOVA technique were also used to understand the statistical behaviour of EMG signals.

REFERENCES

1. N. Uchida, A. Hiraiwa, N. Sonehara, and K. Shimohara, "EMG pattern recognition by neural networks for multi finger control," Proceedings of the Annual International Conference of the Engineering in Medicine and Biology Society, 1992, Vol. 14, pp. 1016–1018.

2. D. Peleg, E. Braiman, E. Yom-Tov, and G.F. Inbar, "Classification of finger activation for use in a robotic prosthesis arm," IEEE Transactions on Neural Systems and Rehabilitation Engineering, 2002, 10(4), 290–293.
3. K. Nagata, K. Adno, K. Magatani, and M. Yamada, "A classification method of hand movements using multi-channel electrode," Proceedings of the 27th Annual International Conference of the Engineering in Medicine and Biology Society, Shanghai, China, 2005, pp. 2375–2378.
4. R. Weir, "Design of artificial arms and hands for prosthetic applications," In Standard Handbook of Biomedical Engineering & Design, M. Kutz, Ed. New York: McGraw-Hill, 2003, pp. 32.1–32.61.
5. K.R. Wheeler, and C.C. Jorgensen, "Gestures as input: neuroelectric joysticks and keyboards," IEEE Pervasive Computing, 2003; 2(2), 57–61.
6. D. Taylor and F. Finley, "Multiple axis prosthesis control by muscle synergies," In Proceedings of the International Symposium on Control of Upper Extremity Prostheses and Orthoses, Göteborg, Sweden, 1971.
7. E. Lamounier, A. Soares, A. Andrade, and R. Carrijo, "A virtual prosthesis control based on neural networks for EMG pattern classification," Artificial Intelligence and Soft Computing, 2002.
8. T.R. Farrell, and R.F. Weir, "A comparison of the effects of electrode implantation and targeting on pattern classification accuracy for prosthesis control," IEEE Transactions on Biomedical Engineering, 2008; 55, 2198–2211.
9. H.J. Hermens, B. Freriks, C. Disselhorst-Klug, et al., "Development of recommendations for SEMG sensors and sensor placement procedures," Journal of Electromyography and Kinesiology, 2000; 10 (5), 361–374.
10. A. Boxtel, P. Goudswaard, and L. Schomaker, "Amplitude and bandwidth of the frontalis surface EMG: effects of electrode parameters," Psychophysiology, 1984; 21 (6), 699–707.
11. R.H. Chowdhury, M.B. Reaz, M.A.B.M. Ali, et al., "Surface electromyography signal processing and classification techniques," Sensors, 2013; 13 (9), 12431–12466.
12. M. Reaz, M. Hussain, and F. Mohd-Yasin, "Techniques of EMG signal analysis: detection, processing, classification and applications," Biological Procedures Online, 2006; 8 (1), 11–35.
13. C. Sapsanis, G. Georgoulas, and A. Tzes, "EMG based classification of basic hand movements based on time-frequency features," In 21st Mediterranean Conference on Control & Automation (MED), pp. 716–722, IEEE, 2013.
14. A. Phinyomark, C. Limsakul, and P. Phukpattaranont, "Application of wavelet analysis in EMG feature extraction for pattern classification," Measurement Science Review, 2011; 11 (2), 45–52.
15. E.L. van den Broek, V. Lis`y, J.H. Janssen, et al., "Affective man–machine interface: unveiling human emotions through biosignals," In Biomedical Engineering Systems and Technologies, pp. 21–47, Springer, 2010.
16. A. Phinyomark, P. Phukpattaranont, and C. Limsakul, "Feature reduction and selection for EMG signal classification," Expert Systems with Applications, 2012; 39 (8), 7420–7431.
17. J.-U. Chu, I. Moon, and M.-S. Mun, "A real-time EMG pattern recognition based on linear–nonlinear feature projection for multifunction myoelectric hand," Proceedings of the IEEE 9th International Conference on Rehabilitation Robotics, Chicago, IL, USA, 2005, pp. 295–298.

18. M. Lei and Z. Wang, "The study advances and prospects of processing surface EMG signal in prosthesis control," Chinese Journal of Medical Instrumentation, 2001, 15(3), 156–160 (in Chinese).
19. K. Englehart, B. Hudgins, P.A. Parker, and M. Stevenson, "Classification of the myoelectric signal using time-frequency based representations," Medical Engineering & Physics, 1999; 21 (6–7), 431–438.
20. C.J. Sherdan, SPSS Version 12.0 for Window Analysis Without Anguish, Australia: John Wiley, 2005.

6 An Intelligent Solution for E-Waste Collection

Vehicle Routing Optimization

Shailender Singh, Malhar Tidke, Mani Sankar Dasgupta, and Srikanta Routroy

Department of Mechanical Engineering, Birla Institute of Technology and Science, Pilani, Rajasthan, India

6.1 INTRODUCTION

Electronics waste (referred to as e-waste or waste electrical and electronic equipment (WEEE)) is classified into five distinct categories based on functionality, material composition, end-of-life attributes, average weight and life span (Ikhlayel, 2018). Although termed as waste, WEEE may contain valuable materials that can be extracted like gold, silver, palladium and many ferrous and non-ferrous metals. Thus WEEE can also be viewed as a resource; the gleaning of WEEE is termed as urban mining. WEEE also contain hazardous substances like lead, mercury, arsenic, cadmium, selenium, hexavalent chromium and some flame retardants that are considered harmful to the environment (Çetinsaya Özkir et al., 2015). These hazardous substances, if not isolated before disposal, can causes serious damage to the biosphere and posing a threat to the health of living beings. Appropriate WEEE management, including collection, segregation, recycling and a safe disposal system, needs to be implemented to mitigate related challenges.

India has introduced various new policies as well as amendments of old policies related to e-waste in 2018, setting a higher level of collection targets for stakeholders. It also specifies an 10% annual increment in the collection target of WEEE generated. The aim of this policy is to collect back 70% of WEEE by 2023 (CPCB, 2018). This is an ambitious target. It requires a robust reverse logistics infrastructure and up-scaling of scientific processing capacity, while at the same time it emphasizes the urgent need for intensive consumer engagement to change the general attitude towards WEEE disposal in our country.

The estimated Indian WEEE production in 2020 was 5.2 million tons, according to a joint study by ASSOCHAM and EY. A previous study also noted that formal recyclers are handling only around 5% of overall e-waste recycled in India,

DOI: 10.1201/9781003181644-6

while the remaining proportion is being recycled by informal recyclers (Khattar et al., 2007). Around 70% of total WEEE generated in India is contributed by households. According to the Central Pollution Control Board (CPCB), there are 312 authorized e-waste dismantlers/recyclers with a total capacity of 0.78 million tons per year as of 2019. The state of Rajasthan has 26 registered recycling units with a total capacity 90,769 metric tons per annum, out of which six recycler units are located in the city of Jaipur, with a capacity of 27,405 MT (CPCB, 2018). The Government of India has mandated that electronic goods producers are solely responsible for the collection and channelling of e-waste after their 'end of life' under the Extended Producer Responsibility (EPR) scheme (Borthakur, 2015).

The producers generally adopt a mixed model to fulfil the legal obligation of EPR. For a selected number of places and/or for a selected time span, producers set up their own collection centres and implement a suitable take-back scheme. At other times, they liaise with third-party collectors called producer responsibility organizations (PRO) for end-of-life e-waste collection (Elia et al., 2019). They may inform clients about the collection of WEEE with a pre-notified schedule. However, the success of such a scheme also depends upon consumers'/end users' behaviour. A facility should be created for consumers to have ready access to information on when, how and where to dispose of WEEE. Generally, authorized collectors facilitate aggregation and maintain transport logistics for e-waste collection; they also may carry out disassembly and segregation for material recovery with full compliance of environmental standards. The profit margins are generally minimal and therefore, one major economic factor for collectors is optimum resources allocation for e-waste collection. They need to decide every day how many vehicles are required for e-waste collection. An improper collection schedule or non-optimum number of vehicles leads to increase in cost as well as increase in greenhouse gas (GHG) emission (Nowakowski et al., 2017). In this study we investigated this issue scientifically and suggested an optimized solution.

Collection of e-waste can be classified into stationary or mobile (Nowakowski et al., 2018). In the stationary mode of collection, collection centres can be set up at appropriate places in residential areas or supermarkets. Containers can also be placed at different places for collection of WEEE in self-service mode (Nowakowski et al., 2020). When the containers are filled, they are shifted to a nearby collection centre. These collection centres can be set up by producers or by independent agencies (Król et al., 2016). Alternatively, container vehicles can be used to pick up e-waste from households or business centrs and gather them at collection centres. The pick-up can be on a fixed schedule or dynamic. The advantage of this method over stationary collection is that pick-up takes place at users' premises and, if possible, at a convenient time for them (Babaee et al., 2020). The collection centres may use Geographic Information Systems (GIS) and optimization models to maximize the amount of e-waste collection from a certain location in a time frame.

The objective of this study was to construct an efficient 'on-demand service' through an intelligent and responsive vehicle routing model. Such a model would benefit the formal collection sector for determination of a waste collection route for multiple vehicles on a daily basis. In order to schedule a pick-up facility for waste disposal, customers can have multiple choices like mobile apps, web-based portals and direct telephonic booking. Data to be recorded are location, time of pick-up, amount and type of waste. The model works on an average delay and maximum delay of waste collection and the same is formulated as a capacitated vehicle routing problem (CVRP). A hybrid evolutionary genetic algorithm is employed which is a combination of a genetic algorithm (GA) and ant colony optimization (ACO). The model can help in reducting collection costs as well as reducing GHG emissions based on optimization of total travel distance.

6.2 PROBLEM DESCRIPTION AND MODELLING

An authorized e-waste collection enterprise named X is situated in Vishwakarma Industrial Area, Jaipur, Rajasthan (Figure 6.1). Its catchment area map is presented in Figure 6.1. It is an authorized PRO and it collects from bulk consumers as well as households. The company has franchises with various producers like Havells, Sony Erikson, Acer, LG, Samsung and Dell for WEEE collection on their behalf. The company wants to implement an 'on-demand service' where customers can book a WEEE collection service by placing an online order. The company owns five vehicles for collection operation and each vehicle has capacity for 25 units;

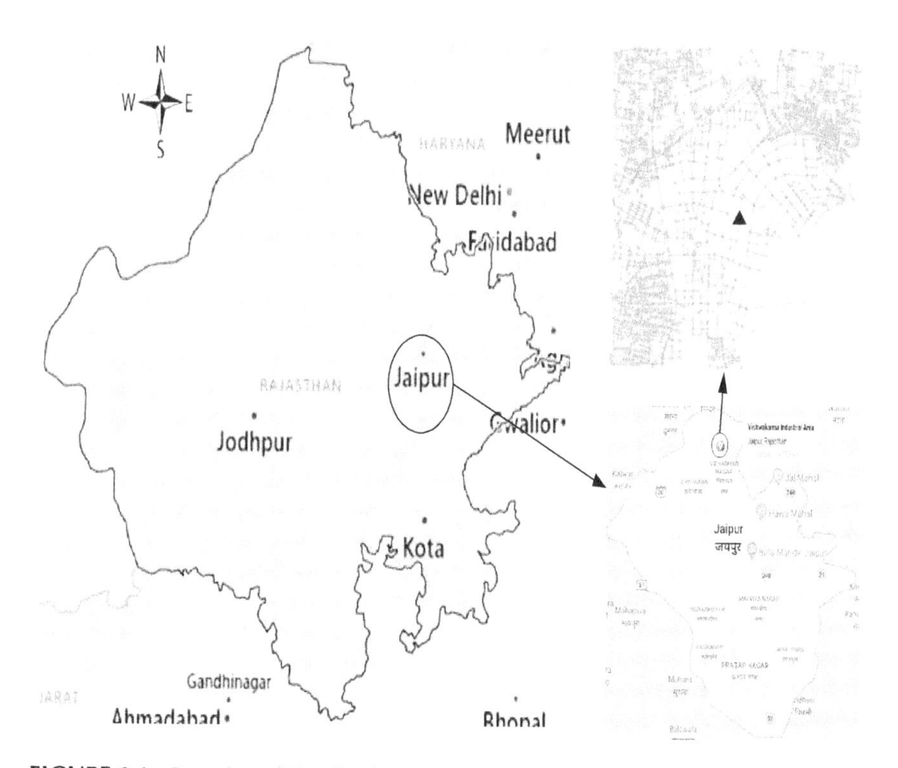

FIGURE 6.1 Location of X collection centre (depot marked with black triangle).

if required, additional vehicles can be hired. One unit of WEEE collectable is assumed to be one-tenth of vehicle capacity. The intelligent scheduling service also resolves the order in which customers are to be serviced based on optimum number of vehicles required. The route for travelling between any two locations is optimized based on the shortest path algorithm and assumptions are made which help in improving productivity and in reducing fuel consumption. The simplifying assumptions are as follows:

- A vehicle is always empty when it leaves the depot.
- The waste from a customer is always completely acquired.
- All customers are always serviced, and time required for servicing and unloading is not a constraint.
- Each customer is always serviced exactly once using only one vehicle and each customer demand is less than truck capacity.
- Volume of waste in the vehicle, at any instance, does not exceed maximum capacity of the vehicle.
- A complete route always starts and ends at the depot.

A reduced-order graphical representation of the problem is shown as in Figure 6.2, where all the serviced customers and the main depot are represented as nodes. Each node is connected to every other node by an edge. Blue nodes represent the customers to be serviced and the black node represents the main depot. The route for a vehicle consists of a collection of edges which start and end with the main depot.

Consider a graph $G = (N, E)$ where N is the set of customers and depot and E is the set of edges connecting the nodes.

$N = \{i | i \in \mathbb{Z} \text{ and } 0 \leq i \leq n\}$ represents that there are n customers. The first element, i.e. {0}, represents the depot.

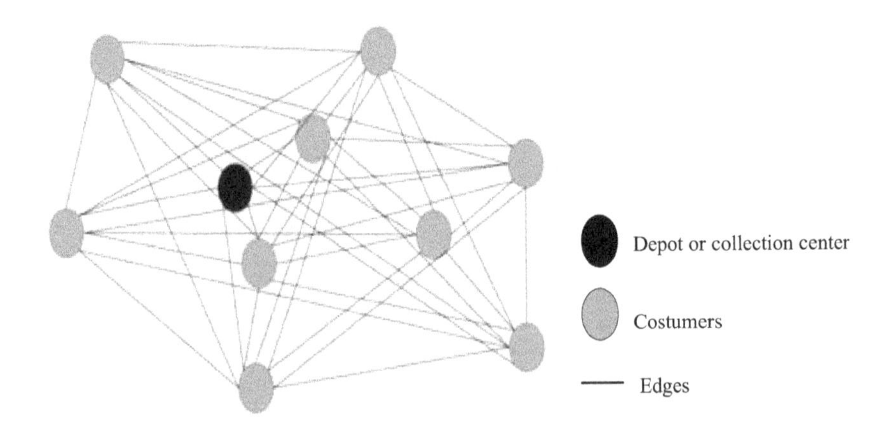

FIGURE 6.2 Representation of the location of customers, depot and edges.

$D = \{d_i \mid \forall\, i,\ d_i \sim \mathcal{N}(\mu, \sigma^2)\}$ represents the demand associated with each customer. Values of demand are taken from a normal distribution and are in m³.

$E = \{(i, j) \mid \forall (i, j) \subseteq N \times N$ and $i \neq j$ and (i, j) are unordered$\}$ represents the set of edges, where each edge is an unordered pair of vertices.

$C = \{C_{ij} \mid \forall (i, j) \in E\}$ represents the set of cost associated with each edge which is taken as the map distance.

$V = \{k \mid k \in \mathbb{Z}$ and $0 \leq k \leq v\}$ represents that v vehicles are used for servicing.

A decision variable x_{ijk} is defined as follows:

$$x_{ijk} = \begin{cases} 1 & \text{if vehicle } k \text{ travels from customer/depot } i \text{ to customer/depot } j \\ 0 & \text{otherwise} \end{cases}$$

Then, the following equation represents objective function for optimization.

$$Z = min \sum_{i=0}^{n} \sum_{j=0,\ j \neq i}^{n} \sum_{k=1}^{v} C_{ij} x_{ijk} \tag{1}$$

The equations for constraints are given as:

- Every customer is visited by only one vehicle only once.

$$\sum_{k=1}^{v} \sum_{j=0,\ j \neq i}^{n} x_{ijk} = 1 \qquad \forall\, i \in N \tag{2}$$

$$\sum_{k=1}^{v} \sum_{i=0,\ i \neq j}^{n} x_{ijk} = 1 \qquad \forall\, j \in N \tag{3}$$

- Volume of waste in each vehicle at a given instance should not exceed the maximum capacity of the vehicle.

$$\sum_{i=0}^{n} d_i \sum_{j=0,\, j \neq i}^{n} x_{ijk} \leq b \qquad \forall\, k \in V \tag{4}$$

- All vehicles always start and end at the depot.

$$\sum_{j=1}^{n} \sum_{k=1}^{k} x_{0jk} = 1 \tag{5}$$

$$\sum_{i=1}^{n} \sum_{k=1}^{k} x_{i0k} = 1 \tag{6}$$

- Decision variable can only be 0 or 1.

$$x_{ijk} \in \{1, 0\} \text{ and } x_{ijk} < \text{truck capacity} \tag{7}$$

6.2.1 Hybrid GA-ACO Algorithm

The proposed algorithm is based on evolutionary theorem and is a hybrid metaheuristic that combined GA and ACO methods (Zukhri and Paputungan, 2013). GA is a population-based method where the initial population is randomly generated. After that, crossover and mutation operations generate a new population. ACO is a stochastic optimization method that imitates the social behaviour of real ant colonies, which try to find the shortest route to feeding sources and back. Real ants lay down quantities of pheromone (a chemical substance) marking the path; the levels of pheromone keep accumulating as the next ant follows and hint towards the most preferred path (Lee et al., 2008). A flow chart of the hybrid algorithm is presented in Figure 6.3.

A conventional GA is employed to calculate the set of parameters of the ACO (Luan et al., 2019). The steps of GA are presented in Figure 6.3; it starts from creation of an initial population of n individuals and then evaluation of the fitness of all n individuals and then loops to check if termination criteria are reached. The function of the crossover operator is to create new sets of child individuals by applying a mathematical function over two individuals of the parent generation. It performs a local search in the vicinity of the two individuals by slightly modifying them. It also helps the algorithm to converge towards an optimum value. Crossover is not initiated for every pair in a generation, but it is controlled by the user through a variable called crossover probability, which is generally set at a high value in the range 0.6–0.8 so that large numbers of crossovers keep happening, although less than 1 to avoid the chance of convergence to a local optima. The function used here is proposed by Deb and Agrawal (1995). The steps followed are:

- Generate a random number $r(0-1)$
- Find the value of α' such that

$$\int_0^{\alpha'} C(\alpha)\,d\alpha = r \qquad if \;\; r < 0.5 \tag{8}$$

$$\int_0^{\alpha'} E(\alpha)\,d\alpha = r - 0.5 \qquad if \;\; r > 0.5 \tag{9}$$

where:

$$C(\alpha) = 0.5(q+1)^{\alpha}$$

$E(\alpha) = 0.5(q+1)\frac{1}{\alpha^{(q+2)}}$ and q is a positive real exponent.

We get the two children individuals Ch_1 and Ch_2 as:

$$Ch_{1,2} = 0.5\left[(Pr_1 + Pr_2) \pm \alpha(Pr_1 - Pr_2)\right] \tag{10}$$

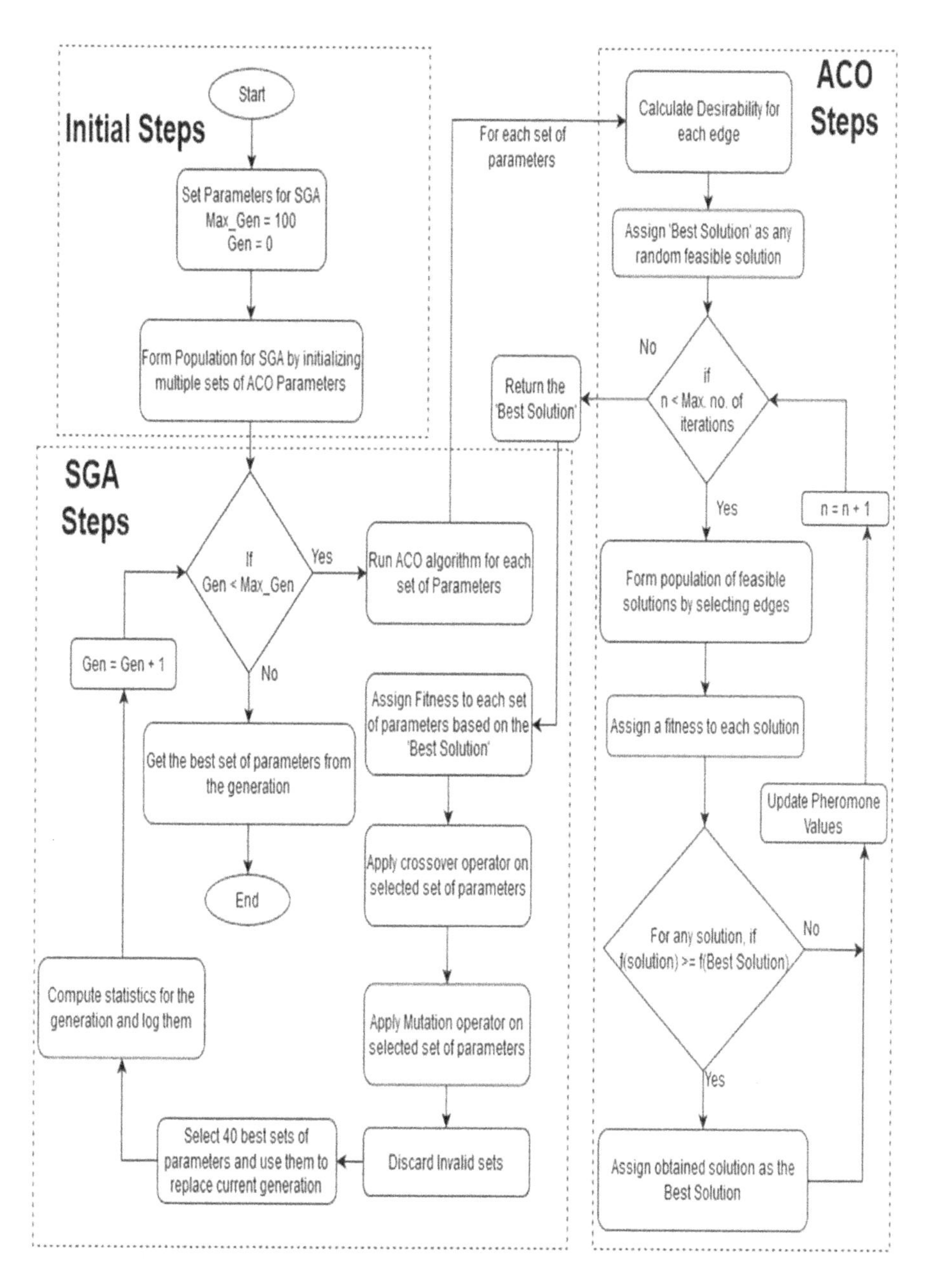

FIGURE 6.3 Flow chart of hybrid algorithm for waste electrical and electronic equipment (WEEE) collection optimization model. ACO, ant colony optimization; and GA, plus genetic algorithm.

The purpose of the mutation operator is to accomplish a global search and to avoid premature loss of genetic material and to maintain diversity in the population. It helps in exploring the maximum of the search space and finding the global optimum. Mutation is, again, a mathematical function implemented on an individual. Higher mutation probability results in a greater number of mutations, many of which may be infeasible solutions and may even cause the search algorithm to digress. Generally it is kept low and in the range 0.2–0.4. The mathematical function implemented for this case study is polynomial mutation, explained as:

- Generate a random number r $(0-1)$.
- Calculate the perturbation factor as:

$$\dot{\delta} = \begin{cases} (2r)^{\left(\frac{1}{q+1}\right)} - 1 & \text{if } r < 0.5 \\ 1 - [2(1-r)]^{\left(\frac{1}{q+1}\right)} & \text{if } r > 0.5 \end{cases} \tag{11}$$

- Mutated solution is given as:

$$Pr_{mutated} = Pr_{original} + \left(\dot{\delta} \times \delta_{max}\right) \tag{12}$$

where δ_{max} user-defined maximum perturbation.

A term penalty is added only for the first five parameters when the values are out of the desired range. The penalty value is calculated as the square of the difference with respect to the upper or lower limit of the range.

$$P_i = (A_i - 0)^2 \quad \text{if } A_i < 0 \tag{13}$$

$$P_i = (A_i - 1)^2 \quad \text{if } A_i > 1 \tag{14}$$

where P_i is the value of penalty and A_i is the value of ith parameter and $\varepsilon[1,5]$. The final value of penalty is then calculated as:

$$Penalty = PenConst \times P_i \tag{15}$$

where the *PenConst* is taken as a very large value of the order of $\sim 10^6$. Fitness function of GA is presented as the following equation:

$$F(parameters) = \frac{L_0}{L + Penalty} \tag{16}$$

Here, L_0 = minimum distance required to cover all the customers.
L = minimum distance achieved by ACO for given set of parameters.

In application of ACO, the individual solution represents a set of routes travelled by vehicles, such that all the customers are serviced and the population is formed by a set of individuals. An edge has three attributes, other than the start node and the end node. ACO as used here is explained in subsequent sections with bullet points.

- Cost (C_{ij}): shortest map distance between the nodes.
- Pheromone (p_{ij}): biologically, pheromones are chemicals which ants secrete to mark a path to a given location. A shorter or frequently used path has a higher density of pheromones and the frequency of the path is considered in the current study.
- Desirability (D_{ij}): desirability of an edge is a combination of pheromones and cost, as presented in following equation.

$$D_{ij} = p_{ij}^{\delta}\left(\frac{1}{C_{ij}}\right)^{\epsilon} \text{ Here, } \delta \text{ and } \varepsilon \text{ are tuning parameters.} \quad (17)$$

It is the value which represents how good an edge is, i.e. if an edge has higher desirability, then it will be selected more often in the best solution. A higher value of δ will tend to increase desirability as it increases the impact of pheromone value and a higher value of ϵ tends to decrease desirability as it increases the impact of edge cost.

- Initial value of pheromones (γ): the initial pheromone value of all edges is equal. Thus, for the first iteration, the selection of edge into the best solution purely depends on its cost.
- Learning rate $\alpha(0 \leq \alpha \leq 1)$: whenever any edge is present in the best individual selected among the population, its pheromone value is increased, as presented in the following equation.

$$p_{ij}^{'} = (1-\alpha)\,p_{ij} + \alpha f(best\ individual) \quad (18)$$

Here, $p_{ij}^{'}$ is the final value of pheromone, p_{ij} is the initial value of pheromone and $f(best\ individual)$ is a function giving the fitness value of the best individual solution based on distance covered by the vehicle in order to complete all scheduled pick-ups. Here, fitness is the ratio of actual distance covered by vehicle for a given solution and minimum distances between any two points, including the central depot, such that all the points are covered.

$$f(individual) = \frac{L_0}{L} \quad (19)$$

Thus, an individual fitness value of 1 is the best individual while the worst individual will have a fitness value of zero. α is the weight given to the ACO fitness value while updating the pheromone value of edges. A higher value of

$\alpha(0.7–1)$ leads to selection of the same edges in every ACO iteration, which further increases its pheromone value and may result in a local optimum. If $\alpha = 0$, then the ACO algorithm turns into a random search algorithm.

- Evaporation rate $\beta(0 \le \beta \le 1)$: to avoid the algorithm becomin stuck in local optima, a higher evaporation rate is introduced, decreasing the pheromone value of all edges with iteration. The values of all pheromones is reduced following equation (20).

$$p_{ij}^{'} = (1-\beta)\, p_{ij} + \beta\gamma \tag{20}$$

A higher value of $\beta\,(0.8–1)$ may lead to selection of different edges in every ACO iteration, resulting in further exploration of solution space looking for a global optimum. If $\beta = 1$, then pheromone value of all edges reduces to γ.

- Desirability tuning parameters δ, ε: the tuning parameter δ is the weight of pheromone value and ϵ is the weight of edge cost in calculating desirability.
- Exploit allowance $(0-1)$: this number helps in deciding which method will be frequently used in edge selection. If the value is higher (0.7-1), then tournament selection will have higher usage compared to the selection based on desirability.
- Population size limit implies a maximum number of individuals in any generation of ACO.
- No. of iterations: number of iterations to be performed for each vehicle.

6.3 SIMULATION

The map and its data for roads and intersections are obtained using GIS from the OpenStreetMap database (OpenStreetMap, 2020). The map is converted into a graph data structure using a library named OSMNX in Python, where the nodes correspond to intersections or terminals and the edges correspond to the roads. Location of the depot is fixed. Location of a node (possible client location) is stored in terms of latitude and longitude in the map and is stored and accessed whenever the shortest path and its length are to be calculated. The demand is assumed to follow a normal distribution and is expressed in terms of volume of WEEE to be collected. Table 6.1 lists the parameters used to generate the number of data calls. μ gives the mean number of calls and σ is the standard deviation. As per Table 6.1, in the first case the number of calls is generated taking 70 as the mean and 3 as the standard deviation.

Table 6.2 gives a sample of data generated; 70 random locations are selected from the map and each of them is assigned a demand.

When a vehicle reaches its limit, it has to return to the depot and unload before attending further scheduled customers. The values of the optimal parameters used in ACO to solve for 110 calls and five vehicles are listed in Table 6.3.The optimal value of GA algorithm is presented in Table 6.4.

TABLE 6.1
Normal distribution parameters for generation of data

Parameters	μ	σ
Number of calls	70 80 90 100 110	3
Demand (m^3)	6	1

TABLE 6.2
Generated data of calls (example)

Customer ID	Latitude	Longitude	Demand (m^3)
1	26.98086	75.77371	4
2	27.00195	75.80179	6
3	26.98947	75.7758	7
4	26.9838	75.76662	6

TABLE 6.3
Value of ant colony optimization (ACO) parameters for N100-V5

Parameters	Values
Learning rate (α)	0.2316
Evaporation rate (β)	0.3179
Tuning parameter corresponding to pheromones (δ)	0.0806
Tuning parameter corresponding to cost (ε)	0.1743
Initial value of pheromones (γ)	0.2879
Exploit allowance	0.8413
Population size limit	20
No. of iterations for a given vehicle	10

6.4 RESULTS AND DISCUSSION

Table 6.5 lists the results of the simulation in terms of fitness values when five vehicles are used for a given number of calls. As the number of calls increase, the total distance travelled to attend all those calls also increases, which results in lowering of fitness values. The optimum number of vehicles required for a given number of calls is determined from the total distance travelled by the vehicles.

TABLE 6.4
Values of genetic algorithm (GA) parameters

Parameters	Values
Number of generations	100
Population size	40
Crossover probability	0.8
Mutation probability	0.2
Exponent in crossover and mutation (q)	5
Maximum perturbation factor (δ max)	0.1
Penalty coefficients	10^6

TABLE 6.5
Maximum fitness value for a given vehicles used and number of calls

No. of vehicles used	No. of calls				
	70	80	90	100	110
1	0.08518	0.08117	0.07632	0.06489	0.059734
2	0.08521	0.08068	0.07685	0.06531	0.059069
3	0.08219	0.07811	0.07704	0.06541	0.05956
4	0.08502	0.07854	0.07622	0.0642	0.061917
5	0.08104	0.07856	0.07591	0.06476	0.060984

However this result will be sensitive to demand per call and total demand in a set of calls, which is assumed to follow a normal distribution. Therefore, to assess the optimum number of vehicles, a range in discrete numbers above and below that is found from simulation, as shown in Table 6.6. The optimized shortest route for a case of four vehicles for 110 calls is presented in Figure 6.4.

Based on the results of the case study, it established that a better optimization method helps in reducing the number of vehicles required on a daily basis for an on-demand service. The algorithm is developed for multiple vehicles and the shortest route identified to improve collection efficiency. Issues such as consumer awareness and data security influence the on-demand collection service. Consumers are always willing to dispose of their WEEE in a convenient way. Any improvement in service will also lead to greater consumer satisfaction with the formal WEEE disposal system and help to reduce the waste going into the informal sector.

6.5 CONCLUSION

Recent amendments in India's e-waste management policy have set new collection targets under the EPR concept for stakeholders. This policy encourages large

TABLE 6.6
Total distance travelled by all the vehicles servicing customers (in km)

	No. of calls				
No. of vehicles used	**70**	**80**	**90**	**100**	**110**
1	119.507	**125.139**	134.402	159.057	172.878
2	**117.637**	131.786	134.221	158.524	177.05
3	126.542	133.875	**132.112**	**157.368**	172.909
4	126.152	133.718	136.577	162.197	**168.076**
5	129.446	131.872	146.461	160.381	169.568
Optimal no. of vehicles required	2	2	3	3	4

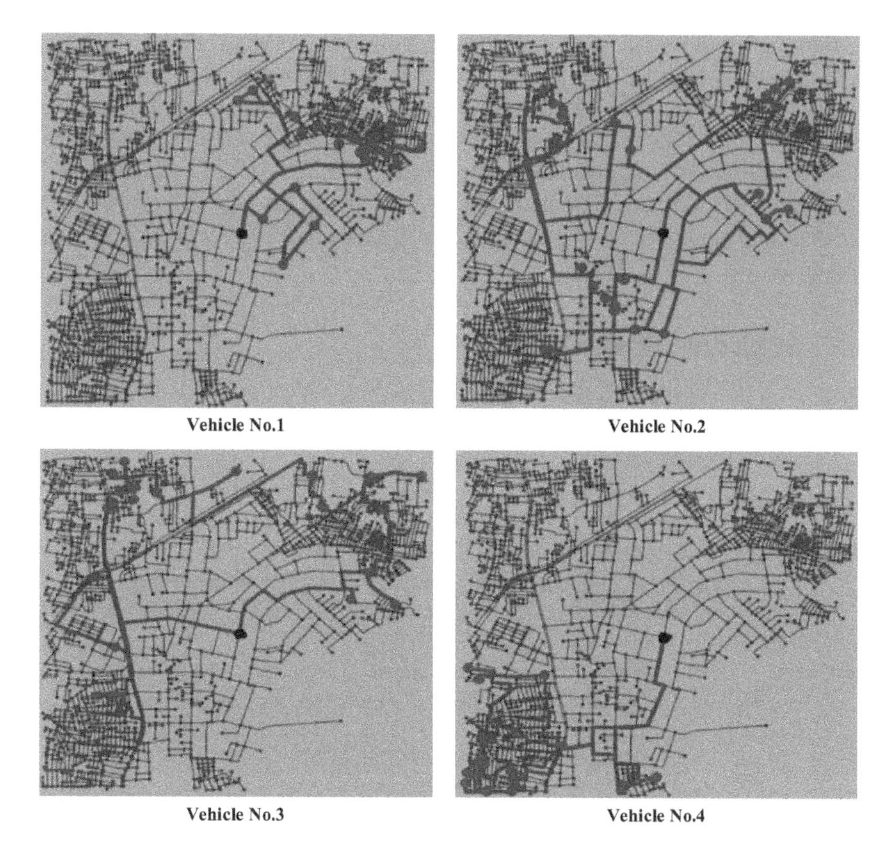

FIGURE 6.4 Optimized routes for vehicles.

business houses producing electrical and electronic equipment to also develop a collection plane for recovery of e-waste from the market. Stakeholders need to implement a sustainable business model as an e-waste collection service for bulk consumers and households and also to assess the demand for e-waste collection services.

This study has presented a model for dynamically scheduling vehicle routes for on-demand WEEE collection based on assumptions, for a real location and fictitious demand data. This type of model can provide better service identifying the shortest path based on the number of calls and demand per call, enhance productivity and reduce fuel consumption as well as reduce GHG emission. The proposed model is an efficient and hybrid algorithm that is flexible and can be applied for any settlement area that can be represented on a digital map. The algorithm is validated using a case study in the city of Jaipur, India and is scalable with the growing size of area. The results obtained show that the model returns optimal routes for collection. Further, recommendations for the optimal number of vehicles can be obtained based on number of calls and their locations, accounting for the capacity of vehicles and workload on each vehicle.

The proposed model is based on evolutionary theorem and can be applied for any demography, with suitable alterations in parametric values. It also ensures more uniform workload distribution among vehicles. Travel time can also be incorporated into the algorithm for future studies. This model can also serve as a foundation for other memetic heuristic algorithms for solving such problems. It can also help companies trying to implement take-back WEEE collection in India to increase their collection.

6.6 ACKNOWLEDGEMENT

This work is partially supported by Department of Science & Technology, Rajasthan, India; grant number P 7(3) Vpro/R&D/2016/3262.

REFERENCES

Babaee Tirkolaee, E., Mahdavi, I., Seyyed Esfahani, M.M. and Weber, G.W., 2020. A hybrid augmented ant colony optimization for the multi-trip capacitated arc routing problem under fuzzy demands for urban solid waste management. *Waste Management & Research*, *38*(2), 156–172.

Borthakur, A., 2015. Generation and management of electronic waste in India: An assessment from stakeholders' perspective. *Journal of Developing Societies*, *31*(2), 220–248.

CPCB, 2018. E-waste (Management) Amendment Rules 2018.

Deb, K. and Agarwal, R., 1995. Simulated binary crossover for continuous search space. *Complex Systems*, 115–148.

Elia, V., Gnoni, M.G. and Tornese, F., 2019. Designing a sustainable dynamic collection service for WEEE: an economic and environmental analysis through simulation. *Waste Management & Research*, *37*(4), 402–411.

Ikhlayel, M., 2018. An integrated approach to establish e-waste management systems for developing countries. *Journal of Cleaner Production*, *170*, 119–130.

Khattar, V., Kaur, J., Chaturvedi, A. and Arora, R., 2007. E-waste assessment in India - Delhi 1–66.

Król, A., Nowakowski, P. and Mrówczyńska, B., 2016. How to improve WEEE management? Novel approach in mobile collection with application of artificial intelligence. *Waste Management*, *50*, 222–233.

Lee, Z.J., Su, S.F., Chuang, C.C. and Liu, K.H., 2008. Genetic algorithm with ant colony optimization (GA-ACO) for multiple sequence alignment. *Applied Soft Computing*, *8*(1), 55–78.

Luan, J., Yao, Z., Zhao, F. and Song, X., 2019. A novel method to solve supplier selection problem: hybrid algorithm of genetic algorithm and ant colony optimization. *Mathematics and Computers in Simulation*, *156*, 294–309.

Nowakowski, P. and Mrówczyńska, B., 2018. Towards sustainable WEEE collection and transportation methods in circular economy – comparative study for rural and urban settlements. *Resources, Conservation and Recycling*, *135*, 93–107.

Nowakowski, P., Król, A. and Mrówczyńska, B., 2017. Supporting mobile WEEE collection on demand: a method for multi-criteria vehicle routing, loading and cost optimisation. *Waste Management*, *69*, 377–392.

Nowakowski, P., Szwarc, K. and Boryczka, U., 2020. Combining an artificial intelligence algorithm and a novel vehicle for sustainable e-waste collection. *Science of the Total Environment*, 138726.

“Open Street Map.” Openstreetmap, www.openstreetmap.org/copyright (accessed 3 September 2020).

Özkır, V.Ç., Efendigil, T., Demirel, T., Demirel, N.C., Deveci, M. and Topçu, B., 2015. A three-stage methodology for initiating an effective management system for electronic waste in Turkey. *Resources, Conservation and Recycling*, *96*, 61–70.

Zukhri, Z. and Paputungan, I.V., 2013. A hybrid optimization algorithm based on genetic algorithm and ant colony optimization. *International Journal of Artificial Intelligence & Applications*, *4*(5), 63.

7 Identification of Most Significant Parameter in Estimation of Solar Irradiance at Any Location

A Review

Shubham Gupta and Amit Kumar Singh
Department of Instrumentation and Control Engineering, Dr. B. R. Ambedkar National Institute of Technology, Jalandhar, India

7.1 INTRODUCTION

Now a days we are focusing on electricity production from clean energy sources. Here the meaning of clean energy sources is those sources which create less pollution in the production of electricity and that are renewal by nature. In India electricity production through conventional methods takes place by the combustion of coal or fossil fuels. Due to combustion of these fuels huge emissions of greenhouse gases occur which were naturally not taken by the plant and source that absorb them. This extra amount of greenhouse gases goes into the atmosphere and traps radiation, and as a result,the temperature of the earth surface increases. Large quantities of green house gas reduce air quality and the air quality index increases or becomes poor. In addition to greenhouse gases there are many sources that increase the National Air Quality Index (NAQI). The sun gives energy in the form of electromagnetic waves, also known as solar energy, and it is assumed that this source will never be depleted. Before establishing a solar farm, we need to ascertain potential solar energy available in that area. Some researchers have predicted that in the year 2050 electricity production through solar photovoltaic (PV) energy will be 22% of total electricity production [1]. This large-scale production of electricity through a solar PV system must have reliability and continuity. The regular supply to load needs a large-scale storage system, which is expensive and requires extensive maintenance. A low-cost

DOI: 10.1201/9781003181644-7

solar PV system is possible if we know everything about the components of generation, transmission and distribution. Then we can optimise the quantity of components following demand, which in turn reduces the cost of the overall system. One of the essential components is solar irradiance. Measurements are recorded of solar radiation parameters using various instruments such as pyrheliometers, pyranometers, pyrradiometers, pyrgemeters, net pyrradiometers and sun photometers. Pyrheliometers are used to measure direct solar irradiance. Pyranometers measure diffuse solar irradiance. Pyrradiometers measure total energy in both solar and terrestrial radiation energy wavelengths. Pyrgemeters measure terrestrial radiation only. Net pyrradiometers measure the balance between total downward and total upward radiant energy or net radiant energy. Sun photometers are used to measure irradiance in narrow-bandpass or interference filters [2]. The accurate and long-term continued measurement and recording of global and diffuse solar radiation are not available at various locations in developing countries, because of high cost and poor facilities for the installation and maintenance of measuring instruments such as pyranometers [3]. In India there are 46 stations under the Indian Metrological Department (IMD) that can measure solar radiation, of which five are defunct, 39 measure global solar radiation (GSR) and 23 measure diffuse solar radiation [2]. So, various methods have been suggested to estimate radiation at any location without measuring radiation. The advantage of this method is reduced cost of installation and maintenance of the measurement set-up. There are many approaches to find GSR and diffuse solar radiation; they can be classified into two categories: single-variable and multi-variable. Here the meaning of single is where only one input variable is used to estimate radiation, whereas in multi-variable approach more than one variable is used to estimate radiation. In an empirical model we generally used the single-variable approach while in soft computing methods we prefer the multi-variable approach.

GSR is a key variable utilized for electricity generation using the PV array. Variations in PV output power are caused by random change (fluctuations) in the irradiance incident on the PV array. The cause of irradiance variation can be divided into two categories: random and deterministic components. The deterministic part is calculated by studying the rotation of earth around the sun and astronomy; some researchers have named it the geographical parameter. Other important and indeterministic fluctuations in PV output power are caused by uncertain behavioural changes in atmospheric parameters. Due to variation in output power voltage and frequency fluctuation occur. To avoid a costly storage system and reliable grid operation we must know in advance the effect of these random variables on PV output power. The random parameters or meteorological variables are cloud cover, wind speed, temperature, visibility, air pressure and humidity level [4, 5]. These meteorological parameters are used in the estimation of solar irradiance. Analysis on the basis of the following parameters is not sufficient to estimate solar radiation because the correlation among the parameters also exists. Without knowing the degree of correlation between the parameters, the most significant parameters responsible for the level of solar radiation at that location cannot be found [6]. The various methods employed to accomplish this task are discussed in the next part of this chapter.

Other parameters which are equally important in an assessment of the production of electrical power by solar panels is amount of dust deposition on solar plates [7]. This parameter is also calculated using an indirect approach. Here the meaning of indirect approach is that the amount of dust deposition can be calculated using pollution data, meteorological parameters and satellite images which are available on different agency websites. Measurement of these data is done for some other purpose and these data are available for analysis without developing the measurement set-up. This approach saves time and money. The accuracy of the solar radiation calculation depends on the soft computing approach. A comparison of soft computing methods can be done with empirical models and/or with another soft computing approach. Here an empirical model is considered as a linear model. The empirical model means that a relationship is established among variables with the help of results and experimental calculations. Some researchers have also used pollution data to predict solar energy at the location of interest. The various soft computing approaches with their advantages and disadvantages are discussed in section 7.2. Our study provides new ideas on the importance of a set of different variables individually (sole) or as a group for estimation of solar irradiance at any location.

7.2 ANN METHODS OF ESTIMATION OF SOLAR ENERGY

The deterministic and stochastic parameters discussed in the introduction to this chapter are responsible for an amount of solar irradiance. Many experimental-based models (empirical) and machine learning methods suggest estimating solar irradiance [8–12]. One of the soft computing methods is artificial neural network (ANN). ANN is a mimic of the human nervous system. ANN is able to solve an unknown input for which it was not trained. The architecture of ANN consists of primarily input layer, connection weight and biases, activation function, summation junction and output layer. Calculation of results using ANN is divided into two stages: learning (training) and recall. The aim of training a network is to update it on weight and bias until the desired accuracy is achieved or there is no further updating of weight. The efficacy of the learning algorithms depends on the number of epochs required to accomplish the desired accuracy. One epoch is defined as one cycle for all the training input. The learning methods are classified on a network's training methods, including supervised, unsupervised, reinforcement and evolutionary learning. Numerous authors have proposed their studies for estimating solar potential using ANN. There are variations in ANN networks due to their different inputs. For estimating daily GSR at any location, the methods and network used include multilayer perception (MLP), radial basis neural network (RBNN), generalized regression neural network and Bienenstock, Cooper and Munro (BCM). Among all the models MLP and RBNN are more accurate compared to others. Values of mean absolute error (MAE) and root mean square error (RMSE) were found to be in the range of 1.53–2.29 and 1.94–3.27 $MJ.m^{-2}day^{-1}$ for the nMLP model.

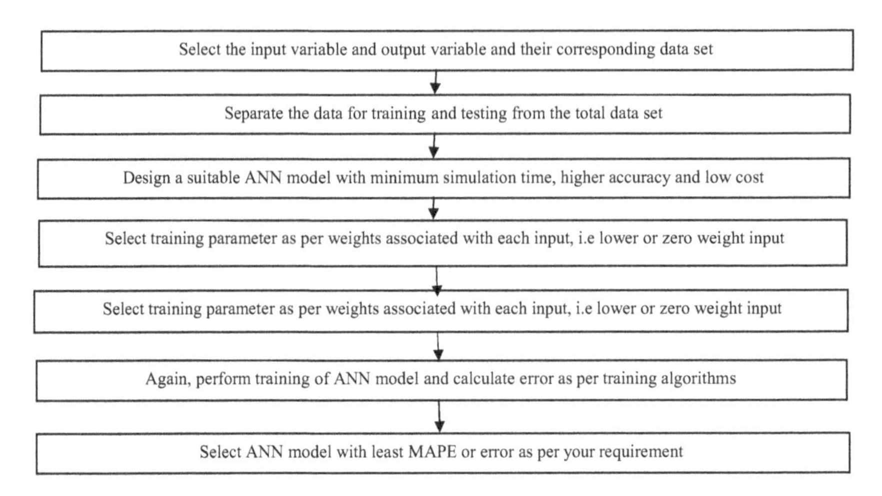

FIGURE 7.1 Flowchart of artificial neural network (ANN) estimation methodology. MAPE, mean absolute percentage error.

Some researchers used MAE, RMSE, mean squared log error (MSLE) and mean bias error (MBE) as measures to check the performance of the model [6]. The above parameters are also used as a measure of compression of regression methods. Several machine learning-based regression methods will be used to estimate PV output power using the meteorological variables discussed [6]. The various regression methods are multivariate linear regression (MLR), LASSO regression (LR), support vector regression and random factor (RF). A flow chart of solar radiation estimation using ANN is shown in Figure 7.1.

7.3 PARAMETERS AND DATA COLLECTION

Parameter, and especially stochastic parameters, can be collected from various sources, including the websites of weather departments, either regional or central, solar energy centres, satellite data from particular country satellite departments, National Aeronautics and Space Administration (NASA) website and pollution data from the pollution department. In India how particular parameter data collection is done is discussed below using the example of pressure; the same method can be applied to find other metrological parameters. One further parameter that may affect solar irradiance is air quality index (AQI). The method of obtaining AQI in India is discussed in section 7.3.2.

7.3.1 Pressure

The atmospheric pressure used to estimate solar irradiance at the location of interest and its data can be collected from regional metrological centres or Mausam Kendra for that region. IMD and its Pune centre provide data on a payment basis; the price depends on who it is buying. Data from the NASA website is free.

There are differences in the magnitude of data from these two departments, IMD and NASA, because of their measurement methods. Data selection depends on the degree of accuracy needed. The pressure of New Delhi Arya Nagar area was collected from the reginal metrological department of Delhi and by applying online to IMD Pune under surface data collection for the corresponding dates. Similar data can be obtained from the NASA website. The steps to get data from the NASA website are as follows:

1. Go to the URL: https://power.larc.nasa.gov
2. Click on the tab power data access viewer.
3. Four small window options are available for data access; choose the one that suits your requirements best.
4. Select power single point data access, choose user community, choose a temporal average, select as per your requirement one of three options, i.e. daily, interannual, climatology.
5. Enter a latitude and longitude for the location or add a point tap on the map.
6. Select the parameters meteorology (moisture and other); then select surface pressure.
7. Submit and get data in a suitable format and download the data.

7.3.2 Air Quality Index

The AQI is one of the measures to check air quality. It shows the amount of dissolved gases and suspended particles in the air. Some researchers uses AQI as a parameter to estimate solar irradiance [13]. The AQI and its constitutent components are data provided by the Central Pollution Control Board (CPCB) of the Government of India by direct contact with its department. An android app SAMEER is also available, developed by CPCB, to enable daily updates on AQI and the amounts of gaseous and particulate matter component. The AQI calendars in the app provide past data on AQI. It is difficult to get data for all locations because CPCB provides data only for locations where monitoring stations exist run by either the CPCB or the state government. The AQI is a very good method; accuracy and advantages have been compared to other methods by Junliang Fan, Lifeng Wu and others using support vector mechanisms [13]. The AQI transforms each air pollution-related parameter into a single number or set of numbers. Transformation takes place with equations or relations that translate parameter values to a simpler form by numerical manipulation. If actual concentrations are reported in $\mu g/m^3$ or ppm (parts per million) along with standard, then it cannot be considered to be an index.

There are two steps involved in formulating an AQI

1. Transformation of each pollutant to sub-indices by a mathematical equation or associated relation to relate X_i to I_i.
2. Accumulation/aggregation of sub-indices to get an AQI.

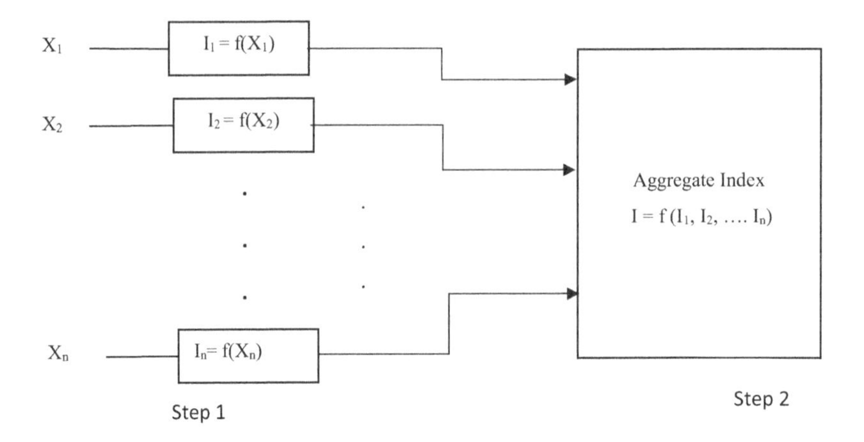

FIGURE 7.2 Formulation of air quality index (AQI).

TABLE 7.1
Parameter standards

Pollutant	SO_2	NO_2	$PM_{2.5}$	PM_{10}	O_3	CO (mg/m³)		Pb	NH_3
Average time (h)	24	24	24	24	1	1	8	24	24
Standard	80	80	60	100	180	180	100	1	100

3. The following methods are adopted in the calculation of sub-indices: weighted additive form, root sum power form (non-linear aggregation form), root mean square form and minimum (min) or maximum (max) operator.
4. The calculation of an aggregate index can be summarized with the help of Figure 7.2. In India, when calculating AQI, we use min or max operator (I = min or max $[I_1, I_2, I_3, \ldots I_n]$).

7.3.2.1 National Air Quality Index in India

In India 12 pollutant parameters are used to decide the NAQI. Of the 12, the first eight parameters (Table 7.1) have short-term (1/8/24 hours) and annual standards (except for CO and O_3), whereas the other four parameters only have annual standards.

Benzo(a)pyrene (BaP), benzene (C_6H_6), arsenic (As) and nickel (Ni) have annual standards. Calculation of NAQI is done on the basis of min or max operator. Nowadays the app information and CPCB AQI bulletin have decreed that Pb has become obsolete or was removed from the analysis.

7.4 LINEAR MODEL OF GSR ESTIMATION

One way to calculate GSR is sunshine hours. This is measured at all locations around the world [14]. Scientists have developed many empirical equations. Some researchers named these models linear models. In an Angstrom-type equation H is monthly average daily global radiation, H_0 is monthly average daily extraterrestrial radiation, S is monthly average daily hours of bright sunshine (h) and S_0 is monthly average day length (h), and a and b are empirical coefficients.

$$\frac{H}{H_0} = \mathrm{a} + \mathrm{b}\frac{S}{S_0} \qquad (1)$$

Model based on equation (1) can be used to relate H and H_0 to different parameters. These parameters are any of the meteorological parameters and other deterministic variables.

$$\frac{H}{H_0} = \mathrm{a} + \mathrm{b}(\frac{RH}{100}) \qquad (2)$$

where RH is relative humidity.

$$\frac{H}{H_0} = \mathrm{a} + \mathrm{b}(T_{\mathrm{avg}}) \qquad (3)$$

A third model is based on precipitation (PT) (mm):

$$\frac{H}{H_0} = a + b(PT) \qquad (4)$$

These models, which relate H and H_0 and its constant values a and b, are calculated using the independent measured data available for that location.

The above-discussed linear model is used for comparison of the accuracy of ANN methods for the same location. The benefit of a linear model is that GSR can be calculated with a smaller number of input or independent variables. Analysis of a linear model is very simple. Equations 2, 3 and 4 were used by researchers to compare the ANN model in the Champa district of Himanchal Pradesh, India [15]. The linear model [16] suggests methods to find empirical constants a and b using the values of latitude (°N), longitude (°E) and altitude (m). An ANN method was used to rank parameters [15]. The ranking was done on the basis of weight associated with each of the input parameters. The order of the magnitude of weight associated with inputs from maximum to minimum are temperature, relative humidity, atmospheric pressure, isolation clearness index, precipitation. The meaning of the ranking here is that the highest-ranking parameter is the most

TABLE 7.2
Range of mean absolute percentage error (MAPE)

Magnitude of MAPE	Prediction accuracy
MAPE ≤ 10%	High
10% ≤ MAPE ≤ 20%	Good
20% ≤ MAPE ≤ 50%	Reasonable
MAPE ≥ 50%	Inaccurate

dominant one; the rest are in order. If the weight is much lighter than others, that parameter can be eliminated from the analysis. This approach reduces the size of the network and simulation time. Many empirical models have been proposed and developed by researchers [17–28]. The prediction is accurate if the measurement and data recorded of a variable are more accurate and can be considered as quality measured data [29, 30]. Researchers used mean absolute percentage error (MAPE) as a measure to judge the performance and its magnitude; performance is shown in Table 7.2 [31].

7.5 CONCLUSION AND FUTURE SCOPE

The above study suggests that the estimation of solar irradiance can take place using variables measured for other purposes and available in various agencies of India. The above study also suggests there was a research gap when estimating better results due to a good NAQI during lockdown in India. A similar approach can be applied to other purposes without data being measured. Soft computing methods improve the accuracy of the estimation of variables under analysis. The advantage of an indirect method is that data are deliberately available for the analysis, and no measurement is needed for the parameter used in the analysis.

REFERENCES

1. IEA, Technology Roadmap Solar Photovoltaic Energy, International Energy Agency (IEA).
2. Tyagi, A.P., (Ed.) Solar Radiant Energy Over India. Indian Meteorological Department, Ministry of Earth Sciences, India; 2009.
3. Jahani, B. and Akhatar, N. Comparison of empirical model to estimate monthly mean diffused solar radiation from measured data: case study for humid-subtropical climatic region of India. Renewable and Sustainable Energy Reviews, 2017; 77: 1326–1342.
4. Raza, M.Q., Nadarajah, M. and Ekanayake, C. On recent advances in PV output power forecast. Solar Energy, 2016; 136: 125e144.
5. Antonanzas, J., Osorio, N., Escobar, R., Urraca, R., Martinez-dePison, F. and Antonanzas-Torres, F. Review of photovoltaic power forecasting. Solar Energy, 2016; 136: 78–111.

6. Alskaif, T., Dev, S., Visser, L. and Hossari, M. A systemic analysis of meteorological variables for PV output power estimation. Renewable Energy, 2020; 163: 12–22.
7. Katazynastyszko, M.J., Teneta, J., Hassan, Q., Burzynska, P., Marcinek, E., Lopain, N. and Samek, L. An analysis of the dust deposition on solar photovoltaic modules. Environmental Science and Pollution Research, 2019; 26: 8393–8401.
8. Wan, K.K., Tang, H.L., Yang, L. et al. An analysis of thermal and solar zone radiation models using an Angstrom–Prescott equation and artificial neural networks. *Energy,* 2008; 33: 1115–1127.
9. Mohandes, M.A. Modeling global solar radiation using particle swarm optimization (PSO). *Solar Energy,* 2012; 86: 3137–3145.
10. Olatomiwa, L., Mekhilef, S., Shamshirband, S. et al. A support vector machine–firefly algorithm-based model for global solar radiation prediction. *Solar Energy,* 2015; 115: 632–644.
11. Feng, L., Qin, W., Wang, L. et al. Comparison of artificial intelligence and physical models for forecasting photosynthetically-active radiation. *Remote Sensing,* 2018; 10: 1855.
12. Wang, L., Hu, B., Kisi, O. et al. Prediction of diffuse photosynthetically active radiation using different soft computing techniques. *Quarterly Journal of the Royal Meteorological Society,* 2017; 143: 2235–2244.
13. Fan, J., Wu, L., Zhang, F., Cai, H., Wang, X., Lu, X. and Xiang, Y. Evaluating the effect of air pollution on global and diffuse solar radiation prediction using support vector machine modeling based on sunshine duration of air temperature. Renewable and Sustainable Energy Reviews, 2018; 94: 732–747.
14. Ahmad, M.J. and Tiwari, G.N. 3. Solar radiation models: review. International Journal of Energy and Environment, 2010; 1(3): 513–532.
15. Kumar, S. and Kaur, T. Efficient solar radiation estimation using cohesive artificial neural network technique with optimal synaptic weights. Journal of Power and Energy, 2019; 0(0): 1–12 IMechE 2019.
16. Kumar, R., Verma, R. and Aggarwal, R.K. Empirical model for the estimation of global solar radiation for Indian locations. *International Journal of Ambient Energy,* 2018; 42: 1–21.
17. Wong, L.T. and Chow, W K. Solar radiation model. Applied Energy, 2001; 69: 191–224.
18. Khatib, T., Mohamed, A. and Sopian, K. A review of solar energy modelling techniques. Renewable and Sustainable Energy Reviews, 2012; 16: 2864–2689.
19. Jebaraj, S. and Iniyan, S. A review of energy models. Renewable and Sustainable Energy Reviews, 2006; 10: 281–311.
20. Sonmete, M.H., Ertekin, C., Menges, H.O., Haciseferogullari, H. and Evrendilek, F. Assessing monthly solar radiation models: a comparative case study in Turkey. Environment Monitoring and Assignment, 2011; 175: 251–77.
21. Batlles, F.J., Rubio, M.A., Tovar, J., Olmo, F.J. and Alados- Arboledas, L. Empirical modelling of hourly direct irradiance by means of hourly global irradiance. Energy, 2000; 25: 657–688.
22. Karakoti, I., Pande, B. and Pandey, K. Evaluation of different diffuse radiation models for Indian stations and predicting the best fit model. Renewable and Sustainable Energy Reviews, 2011; 15: 2378–2384.
23. Karakoti, I., Das, P.K. and Singh, S.K. Predicting monthly mean daily diffuse radiation for India. Applied Energy, 2012; 91: 421–425.

24. Katiyar, A.K. and Pandey, C.K. Simple correlation for estimating the global solar radiation on horizontal surfaces in India. Energy, 2010; 35: 5043–5048.
25. KhoraSanizadeh, H. and Mohammadi, K. Introducing the best model for predicting the monthly mean global solar radiation over six major cities of Iran. Energy, 2013; 51: 257–266.
26. Behrang, M.A., Assareh, E., Noghrehabadi, A.R. and Ghanbarzadeh, A. New sunshine based models for predicting global solar radiation using PSO (particle swarm optimization) technique. Energy, 2011; 36: 3036–3049.
27. Badescu, B., Gueymard, C.A., Chebal, S., Oprea, C., Baciu, M., Dumit, A. et al. Computing global and diffuse solar hourly irradiation on clear sky. Review and testing of 54 models. Renewable and Sustainable Energy Reviews, 2012; 16: 1636–1656.
28. Coskun, C., Oktay, Z. and Dincer, I. Estimation of monthly solar radiation distribution for solar energy system analysis. Energy, 2011; 36: 1319–1323.
29. Myers, D.R. Solar radiation modelling and measurements for renewable energy applications: data and model quality. Energy, 2005; 30: 1517–1531.
30. Muneer, T., Younes, S. and Munawwar, S. Discourses on solar radiation modelling. Renewable and Sustainable Energy Reviews, 2007; 11: 551–602.
31. Lewis, C.D. International and Business Forecasting Methods. London: Butterworths; 1982.

8 Assessment of Sustainable Product Returns and Recovery Practices in Indian Textile Industries

Amit Vishwakarma,[1] M.L. Meena,[1] G.S. Dangayach,[1] and Sumit Gupta[2]

[1] Department of Mechanical Engineering, Malaviya National Institute of Technology Jaipur, India

[2] Department of Mechanical Engineering, ASET, Amity University Noida, India

8.1 INTRODUCTION

To create a future sustainable world, it is necessary that manufacturing industries in India take interest in and help to deliver products that meet sustainability goals and develop sustainable processes. To accomplish this, a few changes have to be implemented in the manufacturing industry with new models and skills. It must be noted that sustainability is not a short-term process; it is a continuously improving, long-term process. Human beings have a fundamental role to play in the cycle of sustainability. With the increasing demand in today's world for resources there is a need to recapture value from unproductive assets resulting from the organization using product returns and recovery practices (PRRP). This is important as organizations neither ignore nor accumulate product returns. There is a need for continuous improvement in setting appropriate strategies and policies to improve PRRP in the manufacturing industry in India.

A product returns and recovery operation is useful practice and is related to the environmental and economic aspects of sustainability. PRRP reduces waste that goes into the environment and provides economic benefits to the organization. Clothing sustainability is widely focused on the 5R concept: reduce, reuse, recycle, redesign and reimagine [1]. In a similar fashion, the 6R concept is a proactive practice that primarily focuses on the improvement of product, process and system levels along supply chains and makes returned

DOI: 10.1201/9781003181644-8

product economically, environmentally and socially viable [2]. It is said that the recovery process is a mixture of 3R concepts [3], dividing recovery into repair, refurbish, remanufacture, cannibalize and recycle. The purpose is to reduce disposal and increase the amount of materials returned to the whole manufacturing life cycle.

The objectives of the present research are to evaluate sustainable product returns and recovery (SPRR) practices in the Indian clothing industry. From this study, it was found that the reuse and remanufacturing of product and material are commonly practised by the textile industry.

8.2 LITERATURE REVIEW

It has been found in the literature that the sustainability concept is applied in three phases in the clothing industry. It starts in the manufacturing phase. Here practices include design for disassembly and eco-friendly clothing. In design for disassembly practice the designer makes the clothes in such a way that the item can be disassembled at the end of its useful life [4]. Some eco-friendly approaches are followed in sustainable garment manufacturing, from raw material selection to the final stage of garment manufacturing [5]. In the second phase sustainability is integrated into supply chain management, mainly using green supply chain management techniques. Some authors have proposed a conceptual framework for demonstrating the influence of green marketing strategy (GMS) on sustainability performance in clothing manufacture [6]. The third stage requires consumers to prefer or buy eco-friendly clothes. So in this way customers play an important role in sustainability. Indirectly they force the manufacturer to manufacture sustainable apparel. Moreover, some practices are common, like identification of all the barriers of clothing sustainability and analysis of the effects of each barrier on social, environmental and economic aspects of sustainability, and then finding the critical barrier among them. The ultimate aim is to eliminate all the barriers [7]. Triple top-line models were applied to identification of the three pillars of sustainability – economy, equity and ecology. They also analysed how these pillars influenced and impacted each other. Across the supply chain economic factors were identified as drivers for the adoption of sustainable practices; the ecology dimension was the weakest pillar [8]. Sustainable manufacturing practices were identified, and product recovery and recycling played an important role in manufacturing sustainability [9].

From the literature various SPRR practices were identified, as shown in Table 8.1.

8.3 RESEARCH METHODOLOGY

The methodology adopted in this research consists of three steps.

1. In the first step a survey questionnaire was prepared. It comprises two parts. In the first part the question is related to the respondent's profile and

TABLE 8.1
Sustainable product returns and recovery (SPRR) practices

S. no.	Code	SPRR practices	Source
1	SPRR1	Reduce resource utilization (energy and water)	[10, 11]
2	SPRR2	Recycling of returned product/material	[12]
3	SPRR3	Reusability of returned product/material	[3]
4	SPRR4	Recovery of returned product/material for further processing	[10]
5	SPRR5	Remanufacturing of returned products as usable product	[2, 8]
6	SPRR6	Redesign of post-use processes and products	[13]

in the second part the question relates to practices related to SPRR on a Likert scale.

2. In the second step a database of Indian textile industries was created. A survey questionnaire was floated in this database of all the apparel companies.
3. In the third step responses were analysed by a statistical tool. Cronbach alpha and *t*-test were performed on the data, and consequently mean and standard were also calculated.

The approach to research carried out by Flynn et al. [14] was followed when conducting the survey research. This study is vital for the clothing manufacturing sector to adopt a sustainability concept. For this, a cross-sectional survey was conducted in the textile industry in India. A five-point Likert-type scale questionnaire was framed for this study.

The survey instrument has two sections. The first section was about the company and respondent profile. In the second section, respondents were asked to rate their level of agreement with factors for SPRR. A total of 152 questionnaires were sent to a pilot study and 32 valid responses were received. A reliability test was applied to examine the consistent results. Cronbach's alpha was used to measure the internal consistency of SPRR practices. A commonly used value for reliability in the literature is 0.70 [15, 16].

8.4 ANALYSIS AND DISCUSSION

The responses were collected and analysed to find out the important practice of SPRR in achieving sustainability. For this study, respondents were asked about SPRR practices. The descriptive analysis of SPRR practices is given in Table 8.2 and shown graphically in Figure 8.1. From Table 8.2 it can clearly be seen that practices that reduce resource utilization (SPRR1) and remanufacturing of returned products as usable product (SPRR5) have high mean values (3.812 and 3.58 respectively); most Indian clothing manufacturing companies emphasize utilization of resources like water and electricity and returned products converted into usable form.

TABLE 8.2
Descriptive statistics of sustainable product returns and recovery (SPRR) practices

SPRR practices	Minimum	Maximum	Mean	Standard deviation
SPRR1	1.0	5.0	3.812	1.0607
SPRR2	1.0	5.0	3.500	1.2181
SPRR3	1.0	5.0	3.500	1.0776
SPRR4	1.0	5.0	3.406	1.0429
SPRR5	1.0	5.0	3.563	1.1426
SPRR6	1.0	5.0	3.281	1.1622

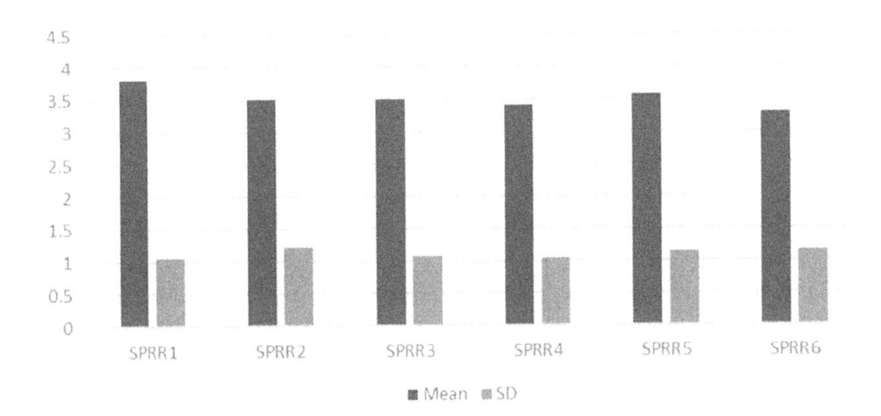

FIGURE 8.1 Descriptive statistics of sustainable product returns and recovery (SPRR) practices. SD, standard deviation.

The bivariate correlation analysis between the SPRR practices is shown in Table 8.3. It can be observed that for both relevant criteria, the correlation is high for all practices. Correlations in Table 8.3 show that Indian clothing manufacturing companies are strongly correlated.

One-sample t-test performed as shown in Table 8.4 identifies the most significant SPRR practice in the clothing manufacturing sector. It is evident that reduce resource utilization (SPRR1) (t = 4.333, 0.000) and remanufacturing of returned products as usable product (SPRR5) (t = 2.738, 0.174) are the most significant practices which are highly adopted by the clothing industry in practice. On the other hand, the remaining SPRR practices are used by companies. As companies put less emphasis on the recovery of returned product/material for further processing (SPRR4) (t = 2.204, 0.035), this means that to achieve sustainability Indian clothing industries should focus on the recovery of returned products.

TABLE 8.3
Correlation between sustainable product returns and recovery (SPRR) practices

SPRR1	SPRR2	SPRP3	SPRR4	SPRR5	SPRR6
1	0.624**	0.677**	0.567**	0.657**	0.743**
0.624**	1	0.811**	0.622**	0.684**	0.638**
0.677**	0.811**	1	0.646**	0.747**	0.747**
0.567**	0.622**	0.646**	1	0.821**	0.684**
0.657**	0.684**	0.747**	0.821**	1	0.727**
0.743**	0.638**	0.747**	0.684**	0.727**	1

** Correlation is significant at the 0.01 level.

TABLE 8.4
One-sample *t*-tests for sustainable product returns and recovery (SPRR) practices

	Test value = 3					
					95% Confidence interval of the difference	
Practices	*t*	df	Significant (two-tailed)	Mean difference	Lower	Upper
SPRR1	4.333	32	0	0.8125	0.43	1.195
SPRR2	2.322	32	0.027	0.5	0.061	0.939
SPRR3	2.625	32	0.013	0.5	0.111	0.889
SPRR4	2.204	32	0.035	0.4063	0.03	0.782
SPRR5	2.738	32	0.01	0.5625	0.143	0.982
SPRR6	1.392	32	0.174	0.2813	–0.131	0.693

8.5 CONCLUSION

SPRR practices are the most important tool for achieving sustainability. This research was performed in Indian clothing industries using a cross-sectional survey. A total of 32 responses were received. The five practices of SPRR were ranked by respondents. The survey responses were analysed by SPSS 22. In this study, it is evident that reduced resource utilization and remanufacturing of returned products as usable product were the most significant practices and recovery of returned product/material was less significant. This study was conducted in the clothing industry; similar research may be carried out in machinery, electrical and electronics and process companies.

REFERENCES

1. Choi, T. M., Lo, C. K., Wong, C. W., Yee, R. W., & Ho, H. P. Y. "A Five-R analysis for sustainable fashion supply chain management in Hong Kong: a case analysis." Journal of Fashion Marketing and Management: An International Journal (2012).
2. Johnson, M. R., & Wang, M. H. "Planning product disassembly for material recovery opportunities." International Journal of Production Research 33, no. 11 (1995): 3119–3142.
3. Thierry, M., Salomon, M., Van Nunen, J., & Van Wassenhove, L. "Strategic issues in product recovery management." California Management Review 37, no. 2 (1995): 114–135
4. Gam, H. J., Cao, H., Bennett, J., Helmkamp, C., & Farr, C. "Application of design for disassembly in men's jacket." International Journal of Clothing Science and Technology (2011).
5. Nayak, R., Panwar, T., & Nguyen, L. V. T. (2020). "Sustainability in fashion and textiles: A survey from developing country." In Sustainable Technologies for Fashion and Textiles (pp. 3–30). Cambridge: Woodhead Publishing.
6. Ara, H., Leen, J. Y. A., & Hassan, S. H. "GMS for Sustainability Performance in the Apparel Manufacturing Industry: A Conceptual Framework." Vision 23, no. 2 (2019): 170–179.
7. Gardas, B. B., Raut, R. D., & Narkhede, B. "Modelling the challenges to sustainability in the textile and apparel (T&A) sector: A Delphi-DEMATEL approach." Sustainable Production and Consumption, 15, (2018): 96–108.
8. Cao, H., Scudder, C., & Dickson, M. A. "Sustainability of apparel supply chain in South Africa: application of the triple top line model." Clothing and Textiles Research Journal, 35, no. 2 (2017): 81–97.
9. Gupta, S., Dangayach, G. S., Singh, A. K., & Rao, P. N. "Analytic hierarchy process (AHP) model for evaluating sustainable manufacturing practices in indian electrical panel industries." Procedia-Social and Behavioral Sciences 189 (2015): 208–216.
10. Ye, F., Zhao, X., Prahinski, C., & Li, Y. "The impact of institutional pressures, top managers' posture and reverse logistics on performance—Evidence from China." International Journal of Production Economics 143, no. 1 (2013): 132–143.
11. Gupta, S., Dangayach, G. S., Singh, A. K., & Rao, P. N. "A Pilot Study of Sustainable Machining Process Design in Indian Process Industry." In CAD/CAM, Robotics and Factories of the Future, pp. 379–385. Springer India, 2016.
12. Gupta, S., & Dangayach, G. S. "Sustainable Waste Management: A Case From Indian Cement Industry." Brazilian Journal of Operations & Production Management 12, no. 2 (2015): 270–279.
13. Gupta, S., Dangayach, G. S., & Singh, A. K. "Key determinants of sustainable product design and manufacturing." Procedia CIRP 26 (2015): 99–102.
14. Flynn, B. B., Sakakibara, S., Schroeder, R. G., Bates, K. A., & Flynn, E. J. "Empirical research methods in operations management." Journal of Operations Management 9, no. 2 (1990): 250–284.
15. Gupta, S., Dangayach, G. S., Singh, A. K., Meena M. L., & Rao, P. N. (2018). Implementation of sustainable manufacturing Practices in Indian manufacturing companies. Benchmarking: An International Journal.
16. Nunnally, J. C. Psychometric Theory. New York: McGraw-Hill (1978).

9 Integrating Reliability-Based Preventive Maintenance in Job Shop Scheduling

A Simulation Study

Shrajal Gupta and Ajai Jain
Department of Mechanical Engineering, National Institute of Technology Kurukshetra, Haryana, India

9.1 INTRODUCTION

In the era of globalization, to fulfill changes in customer expectations, global competition, and other uncertain environment conditions, production planning plays a significant role. Job shop scheduling is one of the most significant aspects of production planning. Job shop scheduling consists of "m" machines, which can simultaneously process "n" job types. Each job requires several operations for completion. Each machine can process only one job at a time (Pinedo, 2008). In a real-time manufacturing scenario, the machine's failure occurs during the processing of jobs as the machine gets old; therefore, there is a need for maintenance activity in scheduling problems. According to different intervals, two types of maintenance activities were considered, i.e., planned maintenance and unplanned maintenance. In unplanned maintenance, a task is executed after a machine fails to restore it to as-good-as-new condition. Planned maintenance symbolizes a set of tasks before machine failure, to keep it in operating condition (Telsang, 1998). A survey by Wang showed seven types of preventive maintenance (PM) policies considered by various researchers. These were periodic PM policy, age-dependent PM policy, reliability-based failure limit policy, sequential PM policy, repair limit policy, repair number computing, and reference time policy (Wang, 2002). The maintenance department does its activities according to its maintenance schedule time. In real-time situations it has been observed that some machines are awaiting maintenance, while jobs are yet to be processed by the machines; this happens due to a lack of coordination between the maintenance and scheduling departments, which affects system performance measures. In this study, an integrated scheduling and maintenance model is considered; the

DOI: 10.1201/9781003181644-9

study provides a recommendation of maintenance timing from the scheduling department to the maintenance department.

9.2 RESEARCH BACKGROUND

In today's manufacturing system, maintenance activities must maintain the system's performance and take care of machine performance. The kind of maintenance approach used in the manufacturing system always affects the performance of the system. Various researchers have considered different types of maintenance approaches to maintain system performance to optimize different types of system performance measures in a scheduling problem. For service production system maintenance Malik (1971) studied age reduction (AR)-based maintenance using improvement factors to adjust the failure rate for imperfect maintenance. He constructed a reliability-based model for the service production system. This maintenance includes functions related to cleaning, lubrication, realignment, and others. He found that the initial risk rate value right after PM changes to hazard rate function because of each imperfect PM.

According to the failure rate, Lie and Chun (1986) proposed a cost-rated model for PM and corrective maintenance (CM). Two types of PM and CM were performed as per schedule and necessity. As introduced by Malik (1971), they considered an improvement factor for failure rate adjustment after maintenance. In this model, the kind of PM and its maintenance cost at the time of PM, and kind of failure and maintenance cost at the time of failure, define which type of corrective action takes place and when. Further, they suggested considering other factors such as maintenance type, average downtime, and average cost during a cycle that affects the improvement factor in future work.

Cassady and Kutanoglu (2003) proposed an integrated model that includes production scheduling and PM planning concurrently. They consider the minimization of the total weight tardiness (TWT) factor as an objective function. The integration model's benefit was demonstrated through a numerical study consisting of a single machine (SM) and three jobs and other parameters, as specified in the paper. Their results showed that there was a 30% improvement in objective function using this method. They suggested considering cost-related objective function, which includes both customer service and failures. In the future, problems related to PM apparatus, labor capability, and multiple machines can be measured.

Zhou et al. (2007) considered condition-based predictive maintenance by combining AR and hazard rate increase method (hazard rate rule) to predict reliability in different maintenance cycles. In their model, imperfect repair activity occurs when systems reach the threshold reliability limit, and once the optimal number of PM cycles was completed, the system is replaced. Their results showed that the optimum reliability threshold (RT) for schedule PM and lowest cost per unit time in the system residual life was 0.93 and 1.8743, respectively. They found that the system's residual life minimizes the cumulative maintenance cost per unit as optimal RT was deduced. Finally, they concluded that the reliability-centered predictive maintenance policy decreased the PM time intervals and was much more practical than the age-T policy.

Liao et al. (2010) studied a reliability-centered sequential PM model considering failure rate function and operation cost. They considered that when the system reached the RT limit, imperfect repair action took place, and once the system reached the optimum number of PM cycles, it was replaced. They accumulated AR factor and failure rate increase factor into failure rate function, which shows more real deteriorating processes. The operational cost meets the requirements of a real situation in this study. The proposed policy performs better than the periodic maintenance policy or age-T policy because their results showed that the proposed policy reduces cost and decreases the time interval for maintenance cycles. It can help meet the goal of near-zero inventories for spare parts. For the future, work can be extended by obtaining improvement factors for system failure rate function.

In contrast to the maintenance model proposed by Liao et al. (2010), Khatab (2018) developed a new model with cost components. His approach considered that there were no restrictions on the decision variable compared to the Liao model (Liao et al., 2010). His work contributes to an understanding of current manufacturing system maintenance by including two costs in the model, i.e., operation and breakdown costs. He suggested a more general repair process instead of minimal repair between PMs. He also suggested a real-time model that consists of a bivariate (age and usage) model to assess the system's actual reliability or condition.

Pan et al. (2010) studied an integrated scheduling model combining both production scheduling and PM planning for an SM manufacturing system, aiming at minimization of maximum weighted tardiness. In their model, they considered flexible time intervals for PM during planning. They considered three cases – scheduling without maintenance, individual with maintenance, and integrated with maintenance – to compare performance. They found that including maintenance in scheduling problems decreases tardiness, enhances the system's efficiency, and saves the cost of malfunction. It also avoids machine idleness. Multiple machines and flow shop scheduling problems can be considered for future work. They suggested that other multi-performance criteria problems, generalized repair problems, maintenance time problems, and problems related to the machine's effects due to jobs should be considered for future research.

Sharma and Jain (2015) monitored flexible job shop scheduling systems with set-up time. The model consists of a dynamic job arrival. They found that flexibility plays a major role in this problem. The reduction in flow time improves overall lead time. Problems related to limited capacity buffer between machines, machine breakdown, batch mode schedule, and external disturbances such as order cancellation and job preemption can be considered future work.

Jamshidi and Esfahani (2015) proposed a mixed-integer non-linear model for scheduling and maintenance problems using a reliability-centered maintenance (RCM) approach. They assumed failure-free time and repair time to be predefined and known in advance for identical parallel machines. They used exponential failure distribution for the problem. They concluded that a decrease in failure rate increased repair cost but reduced quality cost.

Mokhtari and Dadgar (2015) considered time-varying machine failure rate for flexible job shop scheduling problems in a stochastic environment. Mixed-integer linear programming was used to address the problem. To evaluate and assess the availability of shop, the reliability approach was used. Simulated annealing and Monte Carlo simulator approaches were used to minimize the number of tardy jobs (NOTJ). Failure and repair rates were described by exponential probability distribution for each machine. The machine deteriorates at multiple rates and with known rates of repair. The reliability concept was useful in evaluating the shop's availability and determining the best assignment and job sequence on the machine.

Rahmati et al. (2018) considered the stochastic RCM approach with a flexible job shop scheduling problem. Makespan (C_{max}), maintenance cost, and system reliability were the three objective functions in this problem. The PM or CM maintenance actions depend on the system reliability shock degradation level, not by predetermined intervals. The machine degradation and stochastic time between two shocks were described by exponential distribution and maintenance duration as log-normal distribution. This model helps decide which maintenance activity should be done and in a real-time system according to reliability level. Their results showed that the proposed method could control the process intelligently and autonomously. They concluded that the algorithm could monitor and conduct the optimization process based on predetermined RCM consideration efficiently. They suggested using the concept of RCM with more real-world problems. Problems related to redundant machines, as a part of the redundant allocation problem, as well as concurrently balancing the reliability of the system and its cost, can be considered for future research.

Chen et al. (2020) suggested an accurate maintenance model based on reliability intervals under flexible job shop scheduling problems and set-up time. They compared the model with a single RT and periodic maintenance strategy for validation. They found that the optimum RT value was 0.82. They also concluded that their model performed better in maintaining a higher level of machine availability and reliability. Machine learning concepts to refine the parameters of reliability intervals and maintenance time, problems related to random machine breakdown and product features, and problems related to combining real-time production scheduling information and PM can be considered for future research.

From the above literature review the following conclusions can be drawn:

- Most researchers used a single machine and only one objective function problem in their work.
- Most researchers considered two-parameter Weibull distribution for machine failure rate generation. Only a few considered exponential failure distribution.
- The failure rate policy was used with different functions such as AR, hazard rate increase, time-varying function, cost, and failure rate increase function.

- The reliability-based maintenance approach with stochastic technique in real-world problems such as more generalized maintenance and repair time can be studied in future research, as suggested by the researcher (Jamshidi and Esfahani, 2015; Rahmati et al., 2018; Chen et al., 2020).

In summary, these studies deal with single-machine problems. Only a few researchers have considered the RCM approach in scheduling problems; the RCM approach in scheduling problems was based on condition-based maintenance. According to performance measures, the recommendation of PM timing from the scheduling department to the maintenance department was also lacking. These point to motives to integrate a reliability-based PM approach in job shop scheduling problems in a stochastic environment. In the predetermined reliability-centered PM (RCPM) approach, maintenance occurred when system reliability reached the lowest reliability level or reliability threshold. In the present study, ten different machines were considered, each with a different scale and shape parameter. The study provides real-time scheduling scenarios and maintenance timing recommendations, according to system performance measures from the scheduling department.

9.3 JOB SHOP CONFIGURATION

In this study, a job shop manufacturing system with an RCPM approach was considered, in a stochastic environment. The shop consists of ten different machines, which can perform various operations. Each machine has a different scale and shape parameter. The shop has one load/unload station and ten different machines with a local input buffer.

9.3.1 JOB DATA

From job type 1 to job type 6, six different types of job arrive at the manufacturing system dynamically with equal probability. Each job type has a different route and number of operations. Job type 4 has four, job type 1 and 6 has five, and job type 2, 5, and 3 have six operations to perform. The shortest processing time (PT) sequencing rule (highest priority given to a job having the shortest PT for the immediate process) is used to assign the task to the machine.

In this study, Table 9.1 shows the PT and route of each job type. PT of each job type on each machine is assumed to be uniformly distributed and stochastic, and changes according to job type and route.

9.3.2 RELIABILITY-CENTERED PREVENTIVE MAINTENANCE APPROACH

In the RCPM approach, whenever the reliability of the machine reaches the lowest reliability level or is at the reliability threshold, maintenance activity will take place and restores the machine into as-good-as-new condition. The reliability-centered approach determines the specific operation time at different reliability intervals. Various researchers have considered two-parameter Weibull distributions in their

TABLE 9.1
Routing and processing time input data set

	Operations					
Job type	1	2	3	4	5	6
Job type 1	M9, U(5, 6)	M6, U(6, 7)	M10,U(7, 8)	M2, U(6, 7)	M4, U(8, 9)	–
Job type 2	M8, U(8, 9)	M3, U(5, 6)	M5, U(6, 6.5)	M10, U(7, 8)	M1, U(4, 5)	M2, U(7, 7.5)
Job type 3	M7, U(6, 7)	M9, U(3, 3.5)	M3, U(6, 7)	M1, U(3, 4)	M4, U(4, 5)	M6, U(10, 11)
Job type 4	M5, U(4, 5)	M7, U(9, 10)	M9, U(6, 6.5)	M8, U(7, 8)	–	–
Job type 5	M2, U(7, 8)	M1, U(5, 6)	M8, U(6, 6.5)	M10, U(6, 7)	M6, U(4, 4.5)	M4, U(9, 10)
Job type 6	M1, U(10, 11)	M7, U(5, 6)	M9, U(5, 6)	M5, U(9, 10)	M3, U(9, 10)	–

work when studying machine failure (Cassady and Kutanoglu, 2003; Liao et al., 2010; Khatab, 2018; Chen et al., 2020).

Notation

T	Operation time	t	Maintenance time
β	Shape parameter	$\varnothing$	Scale parameter
$R(T)$	Reliability of machine	μPM	Mean value
σPM	Standard deviation	R_s	Reliability threshold

Determining the operation time, which depends on the machine's probability of failure, is closely subjected to Weibull distribution. The failure probability density function g (T) of the machine at operation time T is described as given by Eq. (1) (Cassady and Kutanoglu, 2003; Chen et al., 2020):

$$g(T)=\frac{\beta}{\varnothing}\left(\frac{T}{\varnothing}\right)^{\beta-1}.\exp\left[-\left(\frac{T}{\varnothing}\right)^{\beta}\right],\ T\geq 0 \tag{1}$$

The reliability $R(T)$ of a machine at operation time T can be defined as given by Eq. (2) (Cassady and Kutanoglu, 2003; Chen et al., 2020):

$$R(T)=1-\int_{0}^{T}g(T)\mathrm{dt}=\exp\left[-\left(\frac{T}{\varnothing}\right)^{\beta}\right],\ T\geq 0 \tag{2}$$

The operation time can be calculated as given by Eq. (3).

$$T=(-1)*\varnothing*\left[\ln\left(R(T)\right)\right]^{\frac{1}{\beta}} \tag{3}$$

The literature has observed that, for calculating length of time, i.e., time required for maintenance, normal-log distribution is considered by researchers (Wijaya

TABLE 9.2
Machine parameters

Machine	M1	M2	M3	M4	M5	M6	M7	M8	M9	M10
β	1.6	1.85	1.55	1.7	1.9	1.65	1.95	1.8	1.75	1.5
Ø	62	100	73	89	67	81	94	86	78	90
μ_{PM}	4.02	7.11	3.61	5.92	4.45	4.71	6.67	6.23	5.14	5.56
σ_{PM}	0.107	0.179	0.098	0.152	0.116	0.125	0.17	0.161	0.134	0.143

et al., 2012; Rahmati et al., 2018). Thus, maintenance time is calculated using the following relationship, as given by Eq. (4) (Wijaya et al., 2012; Rahmati et al., 2018).

$$\text{Maintenance time } (t) = \text{log normal } (\mu_{PM}; \sigma_{PM}) \tag{4}$$

where μ_{PM} and σ_{PM} are defined as the mean and standard deviation value of the machine.

In RCPM approach, five different reliability levels i.e. 0.74, 0.78, 0.82, 0.86, and 0.90 are selected (Chen et al., 2020). From Eq. (3), the operation time of a machine or staring time of the maintenance is calculated, with selected reliability levels and machine shape and scale parameters. From Eq. (4), the maintenance time of a machine is calculated with the help of mean and standard deviation value for each machine. The value of β and Ø, and μ_{PM} and σ_{PM} are shown in Table 9.2 for each machine.

9.3.3 Mean Inter-Arrival Time

The mean time between the two jobs is known as mean inter-arrival time. To prevent queues of jobs in front of machines, machine utilization must be less than 100%. Otherwise, queues in front of the machine will grow. Thus, the job inter-arrival time was established using the percentage utilization of the shop and job processing requirements. The job arrival process follows Poisson distribution, as described by Rangsaritratsamee et al. (2004). Thus, the inter-arrival time is exponentially distributed. The mean inter-arrival time of jobs is determined by the following relationship, as given by Eq. (5) (Vinod and Sridharan, 2009; Sharma and Jain, 2015).

$$a = \frac{\mu_p * \mu_g}{S * M} \tag{5}$$

where a = mean inter-arrival time, S = shop utilization, M = number of the machine in the manufacturing system, μ_p = mean PT per operations, and μ_g = the mean number of operations per job.

In this research, $\mu_p = 4.31$ (the mean PT of all operations from Table 9.1). For the input data, μ_g is 5.33 with $M = 10$. Experiments are carried out at 90% shop utilization. From Eq. (5), the relation reveals that the purpose of shop utilization controls the inter-arrival time of jobs. As the number of manufacturing systems increases, the job's inter-arrival time decreases, and shop utilization increases.

9.3.4 Due Date of Jobs

A job order has to be completed on time. The due date of arrival of a particular job is determined either externally or internally. In case of an externally determined due date, this is either fixed by the customer or set for a suitable time in the future. The specific due date is based on the total work content (TWK) at an internally determined due date. TWK represents a sum of the PT of the job or the total number of operations needed on the job. Most researchers use the TWK method to assign the job's due date, as given by Eq. (6) (Vinod and Sridharan, 2011).

$$d_j = a_j + k\,(p_j) \tag{6}$$

where d_j = due date of the job j, a_j = arrival time of the job j, k = due date tightness factor, p_j = total PT of all the operations of job j. In this study, the due date tightness factor (k) = two is considered.

9.4 SIMULATION MODEL CONFIGURATION

The simulation modeling technique is used to analyze and evaluate the system performance manufacturing system. In this study, a discrete-event simulation model for the job shop manufacturing system's operation with a reliability-based PM approach was developed. The job flow in the modeled job shop manufacturing system is shown in Figure 9.1.

The following assumptions are made when developing a simulation model:

- No job can process on more than one machine simultaneously, and any machine can handle any situation.
- The job arrival in the system is dynamic, and a type of job is unknown until it arrives in the system.
- An unlimited capacity buffer is considered before each machine, and pre-emption is not allowed.
- After maintenance of the machine, it is restored to an as-good-as-new state.
- Set-up and transportation time are negligible.
- The scale and shape parameters of each machine are known.

9.4.1 Performance Measures

The system performance measure used in this study for the evaluation purpose was described by Sharma and Jain (2015), as follows:

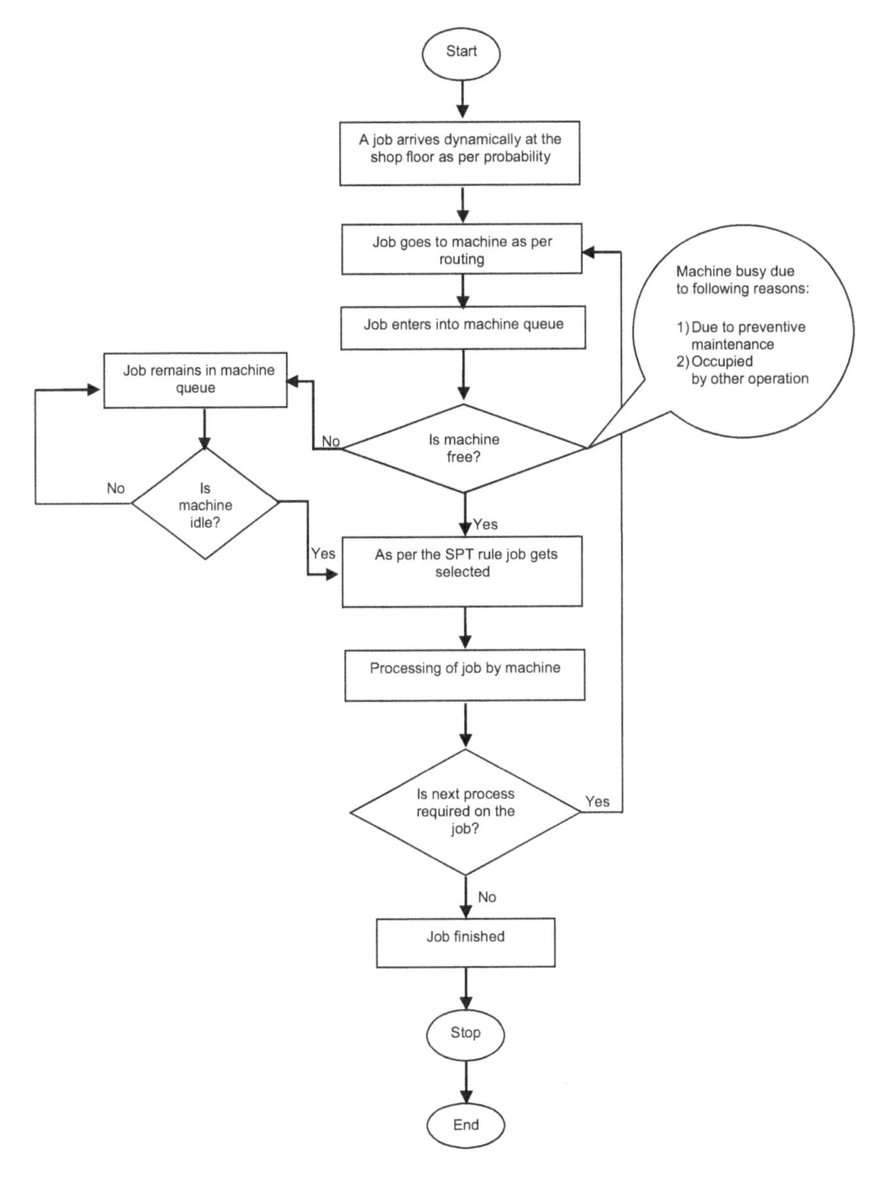

FIGURE 9.1 Flowchart of a job flow. SPT = shortest processing time.

1. C_{max}: defined as the time to finish the last job in a manufacturing system.
2. Mean flow time (MFT): the average time a job spends in a manufacturing system during processing, as given by Eq. (7):

$$\text{MFT} = \frac{1}{n}\sum_{j=1}^{n} F_j \tag{7}$$

Here, $F_j = C_j - a_j$, F_j = the flow time of job j, C_j = the completion time of job j, a_j = the arrival time of job j, and n = the number of jobs produced during the simulation period (during the steady-state period).

3. Mean tardiness (MT): the average delay during processing in the job of a manufacturing system, as given by Eq. (8).

$$\text{MT} = \frac{1}{n}\sum_{j=1}^{n} T_j \tag{8}$$

Here, $T_j = \max\{0, L_j\}$, $L_j = c_j - d_j$, T_j = the tardiness of the job j, L_j = the lateness of job j, d_j = the due date of job j.

4. The NOTJ: The value of the number of jobs completed after their due dates, as given by Eq. (9).

$$\text{NOTJ} = \sum_{j=1}^{n} \delta\left(J_j\right) \tag{9}$$

Here, $\delta(J_j) = 1$ if $J_j > 0$ and $\delta(J_j) = 0$, otherwise.

9.5 EXPERIMENTAL DESIGN FOR A SIMULATION STUDY

In this work, various experiments have been conducted considering two reliability-based maintenance approaches in the job shop manufacturing system with sequence-depeandent set-up time (SDST) using simulation modeling. The first step is to find a transient phase, known as a steady-state period of the simulation experimentation. Welch's procedure, as described by Law and Kelton (2000), was used for a steady-state purpose. A pilot study was conducted for the problem considered. The study revealed that after 250 jobs, the manufacturing system approaches steady state.

For simulation experimentation, ten replications were considered. For the completion of 20,250 jobs, the simulation model ran in each replication. From the simulation output, jobs numbering 1–250 were abandoned due to the transient period, and the output of the remaining 20,000 (jobs numbered 251–20,250) completion jobs was used to measure the performance of the manufacturing system using different performance measures. Table 9.3 shows the layout of the simulation experiments; this table shows all the input parameters used in the experimental problem, including maintenance data.

9.6 SIMULATION RESULTS AND ANALYSIS

In this research, an reliability-centered RCPM approach was carried out using a simulation model for considered job shop scheduling problems. The value of the shop performance measure was shown in Tables 9.4 and 9.5.

TABLE 9.3
Layout of simulation experiments

Machine parameters	Number of machines	10
	Shop utilization	90%
	Scale parameter	62–100
	Shape parameter	1.5–1.95
	Mean value	3.61–7.11
	Standard deviation	0.09–0.179
Job parameters	Job types	6
	Number of operations	04–06
	Mean inter-arrival time	Exponential distribution
	Processing times	Stochastic
	Due date	Total work content method
Maintenance parameter	Reliability levels	0.74, 0.78, 0.82, 0.86, and 0.90
Simulation of job shop	Sequencing rule	Shortest processing time
	Number of replications	10
	Run-length	20,000 jobs completion (after a warm-up period of 250 jobs)

TABLE 9.4
Results of reliability-centered preventive maintenance

R_s	MFT (min)	C_{max} (min)	MT	NOTJ
0.74	214799.86	440277.4	0.000764	0.67
0.78	223777.60	456798.0	0.000478	0.56
0.82	235489.60	481368.5	0.000335	0.29
0.86	255809.40	521015.0	2566.99	820.65
0.90	289598.20	587908.3	39241.43	3024.37

R_s, reliability threshold; MFT, mean flow time; C_{max}, makespan; MT, mean tardiness; NOTJ, number of tardy jobs.

TABLE 9.5
Percentage change in the performance measure

Reliability threshold	w.r.t. reliability threshold	Percentage change			
		MFT (min)	C_{max} (min)	MT	NOTJ
0.78	0.74	4.18	3.75	–37.43	–16.2
0.82	0.78	5.23	5.38	–29.92	–48.94
0.86	0.82	8.63	8.24	7.66E+08	2.82E+05
0.90	0.86	13.21	12.84	1428.69	268.53

MFT, mean flow time; C_{max}, makespan; MT, mean tardiness; NOTJ, number of tardy jobs.

Table 9.4 presents the system's performance as measured by MFT, C_{max}, MT, and NOTJ as each machine's reliability level varied from 0.74 to 0.90 in steps of 0.04 for the job shop considered. Table 9.5 presents a percentage change in the performance measure as the reliability level varied from 0.74 to 0.90 in steps of 0.04. The observations drawn from these results are as follows:

1. Table 9.4 indicates that as the reliability level increases from 0.74 to 0.90, the MFT increases from 214799.86 to 289598.20, and C_{max} increases from 440277.40 to 587908.3. This change occurs because as the reliability level increases, the time between two maintenances decreases. Thus, the job remains on the shop floor for a longer period of time and waits for the machine for processing. This results in higher MFT and C_{max}.
2. Table 9.5 indicates that for MFT performance measure, as the reliability level increases to 0.78 from 0.74, MFT increases by 4.18%. Further, as the reliability level is increased to 0.82 from 0.78, the MFT increases by 5.23%. Also, with an increase in reliability level to 0.86 from 0.82, the MFT increased by 8.63%. As the reliability level is increased to 0.90 from 0.86, the MFT rises by 13.21%. This indicates that at lower levels of reliability, i.e., 0.74, 0.78, 0.82, the change in MFT is about 5% or less. At higher reliability levels, i.e., 0.86 and 0.90, the MFT difference is about 13% or less. Thus, from the MFT viewpoint, lower levels of reliability are suggested.
3. Table 9.5 also indicates that for C_{max} performance measures, as the reliability level increases from 0.74 to 0.78, C_{max} increased by 3.75%. Further, with an increase in reliability level to 0.82 from 0.78, C_{max} increased by 5.38%. With the increase in reliability level to 0.86 from 0.82 C_{max} increases by 8.24%. Further, as the reliability level increases to 0.90 from 0.86, C_{max} increased by 12.84%. This indicates that at lower levels of reliability, i.e., 0.74, 0.78, and 0.82, the change in C_{max} performance measure is about 5% or less. At higher reliability levels, i.e., 0.86 and 0.90, the increase in C_{max} is about 13% or less. Thus, from the C_{max} performance measure viewpoint, lower levels of reliability are suggested.
4. Table 9.4 indicates that as the reliability level is increased from 0.74 to 0.82, MT decreases from 0.000764 to 0.000335, and NOTJ decreases from 0.67 to 0.29. But, as the reliability level increases from 0.86 to 0.90, MT increases from 2566.99 to 39241.3, and NOTJ increases from 820.56 to 3024.37. This change occurs because as the reliability level increases, the time between two maintenances decreases. Thus, the job remains on the shop floor for a longer period and waits for the machine for processing. This results in a change in MT and NOTJ.
5. Table 9.5 indicates that for MT performance measure, as the reliability level increases to 0.78 from 0.74, MT decreases by 37.43%. Further, as the reliability level is increased to 0.82 from 0.78, MT decreased by 29.92%. But, as the reliability level increases from 0.82 to 0.86, MT increases by 7.66×10^8. Further, as the reliability level is increased to 0.90 from 0.86, MT increases by 1.4×10^3. This indicates that at 0.82 reliability level, MT

is lowest. As reliability increases from 0.82 to 0.90, MT increases substantially, which shows that MT is very sensitive to the higher reliability level; even one level shift of reliability level increases MT substantially. Thus, from the MT viewpoint, the optimum reliability level, i.e., 0.82, is suggested.

6. Table 9.5 indicates that for NOTJ performance measure, as the reliability level increases to 0.78 from 0.74, NOTJ decreases by 16.20%. Further, with the increase in reliability level to 0.82 from 0.78, the NOTJ decreases by 48.94%. But, as the reliability level increases from 0.82 to 0.86, NOTJ increased by 2.82×10^5. Further, as the reliability level is increased to 0.90 from 0.86, the NOTJ increases by 268.53%. This indicates that at 0.82 reliability level, NOTJ is lowest. As reliability increases from 0.82 to 0.90, NOTJ increases substantially, which shows that NOTJ is very sensitive to the higher reliability level, and even one level shift of reliability level increases NOTJ substantially. Thus, from the NOTJ viewpoint, the optimum reliability level, i.e., 0.82, is suggested.

9.7 CONCLUSIONS

In this study, the RCPM approach was evaluated for job shop scheduling problems. The shop consists of ten different machines and six job types. The performance of the manufacturing system was measured with the help of four performance measures. The simulation study was conducted using Pro-Model software. According to the results of the maintenance approach the following conclusions were drawn:

1. Lower levels of reliability, viz. 0.74, 0.78, and 0.82, are recommended for MFT and C_{max} performance measures.
2. For due date performance measure, i.e., MT and NOTJ, a 0.82 reliability level is recommended.
3. As the reliability level increases, the percentage change in MFT and C_{max} increases by the same percentage rate.
4. MT and NOTJ performance measures are very sensitive to higher reliability levels, i.e., above the 0.82 reliability level. Even with one level shift of reliability level, these two performance measures increase substantially.

The present research can be furthered by considering more real-time conditions in a scheduling problem. Future research could be directed towards involving situations such as including more PM policies, transportation time, set-up time, and external disturbances such as order cancellation and job preemption.

REFERENCES

Cassady, C. R., & Kutanoglu, E., 2003. Minimizing job tardiness using integrated preventive maintenance planning and production scheduling. *IIE Transactions* (*Institute of Industrial Engineers*), *35*(6), 503–513. https://doi.org/10.1080/07408170304416

Chen, X., An, Y., Zhang, Z., & Li, Y., 2020. An approximate nondominated sorting genetic algorithm to integrate optimization of production scheduling and accurate maintenance based on reliability intervals. *Journal of Manufacturing Systems*, *54*, 227–241. https://doi.org/10.1016/j.jmsy.2019.12.004

Jamshidi, R., & Esfahani, M. M. S., 2015. Reliability-based maintenance and job scheduling for identical parallel machines. *International Journal of Production Research*, *53*(4), 1216–1227. https://doi.org/10.1080/00207543.2014.951739

Khatab, A., 2018. Maintenance optimization in failure-prone systems under imperfect preventive maintenance. *Journal of Intelligent Manufacturing*, *29*(3), 707–717. https://doi.org/10.1007/s10845-018-1390-2

Law, A. M., & Kelton, W. D., 2000. Simulation Modeling and Analysis. Singapore: McGraw-Hill.

Liao, W., Pan, E., & Xi, L., 2010. Preventive maintenance scheduling for repairable system with deterioration. *Journal of Intelligent Manufacturing*, *21*(6), 875–884. https://doi.org/10.1007/s10845-009-0264-z

Lie, C. H., & Chun, Y. H., 1986. An algorithm for preventive maintenance policy. *IEEE Transactions on Reliability*, *35*(1), 71–75. https://doi.org/10.1109/TR.1986.4335352

Malik, M. A. K., 1971. Optimal preventive maintenance scheduling. *ASME Pap 71-PEM-4*, 11(3), 221–228.

Mokhtari, H., & Dadgar, M., 2015. Scheduling optimization of a stochastic flexible job-shop system with time-varying machine failure rate. *Computers and Operations Research*, *61*, 31–45. https://doi.org/10.1016/j.cor.2015.02.014

Pan, E., Liao, W., & Xi, L., 2010. Single-machine-based production scheduling model integrated preventive maintenance planning. *International Journal of Advanced Manufacturing Technology*, *50*(1–4), 365–375. https://doi.org/10.1007/s00170-009-2514-9

Pinedo, M. L., 2008. Scheduling: Theory, Algorithms, and Systems. New York: Springer.

Pinjing, H., Fan, L., Hua, Z., & Liming, S., 2013. Recent Developments in the Area of Waste as a Resource, with Particular Reference to the Circular Economy as a Guiding Principle 144–161. https://doi.org/10.1039/9781849737883-00144

Rahmati, S. H. A., Ahmadi, A., & Karimi, B., 2018. Multi-objective evolutionary simulation based optimization mechanism for a novel stochastic reliability centered maintenance problem. *Swarm and Evolutionary Computation*, *40*(February), 255–271. https://doi.org/10.1016/j.swevo.2018.02.010

Rangsaritratsamee, R., Ferrell, W. G., & Kurz, M. B., 2004. Dynamic rescheduling that simultaneously considers efficiency and stability. *Computers and Industrial Engineering*, *46*(1), 1–15.

Sharma, P., & Jain, A. 2015. Effect of routing flexibility and sequencing rules on performance of stochastic flexible job shop manufacturing system with setup times: simulation approach. *Proceedings of International Mechanical Engineers Part B:Journal of Engineering Manufacturing*, *231*(2), 329–345. https://doi.org/10.1177/0954405415576060

Telsang, M., 1998. Industrial Engineering and Production Management. New Delhi: S. Chand.

Vinod, V., & Sridharan, R., 2009. Simulation-based metamodels for scheduling a dynamic job shop with sequence-dependent setup times. *International Journal of Production Research*, *47*(6), 1425–1447. https://doi.org/10.1080/00207540701486082

Vinod, V., & Sridharan, R., 2011. Simulation modeling and analysis of due-date assignment methods and scheduling decision rules in a dynamic job shop production system. *International Journal of Production Economics*, *129*(1), 127–146. https://doi.org/10.1016/j.ijpe.2010.08.017

Wang, H., 2002. A survey of maintenance policies of deteriorating systems. *European Journal of Operational Research*, *139*(3), 469–489. https://doi.org/10.1016/S0377-2217(01)00197-7

Wijaya, A. R., Lundberg, J., & Kumar, U., 2012. Downtime analysis of a scaling machine. *International Journal of Mining, Reclamation and Environment*, *26*(3), 244–260. https://doi.org/10.1080/17480930.2011.603515

Zhou, X., Xi, L., & Lee, J., 2007. Reliability-centered predictive maintenance scheduling for a continuously monitored system subject to degradation. *Reliability Engineering and System Safety*, *92*(4), 530–534. https://doi.org/10.1016/j.ress.2006.01.006

10 Prioritizing Circular Economy Performance Measures

A Case of Indian Rubber Industries

Somesh Agarwal, Mohit Tyagi, and R.K. Garg
Department of Industrial and Production Engineering, Dr. B.R. Ambedkar National Institute of Technology, Jalandhar, Punjab, India

10.1 INTRODUCTION

As the country's population and economic structure are increasing day by day, product-procuring capability and requirements are also increasing, encouraging goods production and in turn waste growth. Waste development and its assimilation are greatly concerning issues in the current scenario. Treatment of waste and its transformation into a useful form are cutting-edge research areas these days, when environmental issue concerns are rapidly rising. The increase in production exploits numerous natural resources and utilizes extensive transportation. It is the main reason for the depletion of reservoirs of precious natural resources due to the imbalance between natural resources excavation and self-regeneration.

People purchase products, use them to their limits, but at the end of the product's life cycle the object is discarded as waste. Waste which could be helpful in another from is only enlarging the huge litter mountain. Waste can be emission waste, solid waste, or liquid waste (Morseletto, 2020). As the environmental problem is accumulating day by day, society wants reform in order to decrease the adverse environmental hazards. Demand for new techniques/philosophies is urged when the present environmental condition is seen to affect human life in the present day, and is set to become the overriding problem in the time ahead (Suhi et al., 2019). A new emerging topic in recent research is the circular economy (CE), which works on environmental sustainability, where the same job reduces waste and offers a sustainable future (Kirchherr et al., 2017). CE is based on the principle of reuse, recycle, and regeneration of the material/products (Ethirajan and Kandasamy, 2019). The CE paradigm works on each type of waste and aims to minimize them, transiting towards a zero-waste economy. The main idea

DOI: 10.1201/9781003181644-10

behind the CE is to maximize the usage characteristics of the product/material. CE is an idea to reduce waste growth by creating a working environment for reutilizing the end of discarded products.

The traditional manufacturing process utilizes the well-established techniques and methods which have been in use for a long time. Industrial revolutions have always created prolonged improvements in the manufacturing industry, and conjointly it has undergone much refashioning (Esposito et al., 2018). The industries have encountered various tools and techniques for increasing production and decreasing wastage within industry processes (Tyagi et al., 2015a). This has influenced the concept of bulk production, or it can instead be stated as overproduction. To create a single product, many natural resources are utilized, and this decreases the treasury of environmental resources (Tyagi et al., 2015b). These ramifications hurt the earth's peace and equilibrium, giving rise to extreme issues. Global warming, natural calamities, pollution, and so forth, are created by these kinds of human activity (Mangla et al., 2018).

A product follows various paths from design to end delivery, including enormous usage of materials and resources. In the beginning, raw material procurement can be either a purely natural resource or a polymer (human-made). In both cases, natural resources are used and are continually being withdrawn from the natural reservoir. This raw material is then transported to the manufacturing industries that produce the semi-finished product or finished product. One stage is increased for industries producing semi-finished products. The transportation of material causes many emissions from fuel and utilizes a large number of human resources.

Additionally, packaging material is brought into play to packing the product/material. Then this manufactured product is distributed to warehouses or distributors, again using high-quality packaging. In the end, the product is delivered to the customer with an additional amount of packaging. The customer uses the product, and at the end of its life cycle, the product is dumped as waste. In this way, it can be seen that there is much waste occurring. Waste is mainly emissions that occurred due to fuel being burned during the transportation of material, packaging waste at various levels, emissions from the fuel-based manufacturing machine and chemicals used, human resources, and several manufacturing tools for a single-use end-product. Ultimately the process suggests that natural resources that have undergone several processes finally are dumped as waste.

CE aims to decrease this wastage no matter what type it is. According to the CE paradigm, if one or more levels from the above-stated production level have been eliminated, waste growth will automatically decrease. If the end use product is again treated as raw material for the manufacturing industry, i.e., waste is treated as a resource, then the extraction of natural resources can be minimized. If the end product can be used directly in the same or another form, then the product's many manufacturing processes can be eliminated.

Additionally, transportation and workforce usage will be minimized. In that manner, CE aids to minimize product cost, decrease natural resource extraction, minimize energy usage and waste growth, and reduce emissions from transportation and manufacturing machines. CE enhances the production capability of the industry and increases customer satisfaction.

It is well known that rubber products have had an important place in society. But motivation to reuse and recycle rubber products is very limited. In many cases rubber materials are mixed with other agents to improve their strength, which creates difficulty in their remanufacturing capability. In most cases, to enhance the strength of tires, rubber is mixed with steel fibers, which hinders its recycling characteristics. The main problem with disposed rubber is that it creates hazards when incinerated. The emission of greenhouse gaseous and polluting materials causes danger to humanity. Mostly automotive rubber recycling rates are quite high but they are significantly narrower when compared to their production capacity. The CE aims to reuse even a single particle to create a zero-waste economy. In the case of rubber industry practices CE can contribute to overcoming the challenges and aids in providing a successful roadmap. For those reasons rubber industries were studied in this research.

This research demonstrates a framework for analyzing CE's performance measures by taking cases from rubber manufacturing industries situated in Punjab, India. Rubber industries are specifically preferred as rubber causes much pollution at the end of its life cycle on either burning or dumping, although if it could be reused for the same or a different purpose, it would be very advantageous for society and the environment. This research has addressed two fundamental objectives:

1. Identification of CE performance from the available literature and interaction with concerned authorities from rubber industries situated in Punjab.
2. Ranking of CE performance measures, finding an interrelation between the parameters providing a generalized solution to industries concerned.

10.2 LITERATURE SURVEY

CE is a trending scheme that attempts to save the environment from hazards and generates monetary benefits to industry. The agenda of CE to protect natural resources for the future prevents extinction of resources, which is a great challenge for government bodies. CE is the opposite of linear economy. The linear economy focuses on extraction, production, and consumption without worrying about natural resources; in contrast, CE tries to regenerate natural resources by reusing used products and materials. This research topic has taken leading researchers' attention, and its popularity is increasing day by day. Many researchers have elaborated on the concept of CE and provided a detailed vision for this paradigm. Kirchherr et al. (2017) proposed 114 definitions of CE on 17 dimensions.

Similarly, Farooque et al. (2019) classify various terminologies related to supply chain sustainability and conceptualize a unifying definition of CE and circular supply chain management. Murray et al. (2017) have done an in-depth study of CE and analyzed its application to and implementation in a sustainable process. Ghisellini et al. (2016) stated that CE approaches could withhold products' consumption, integrated parts, components, or the materials present in each part of the product. Additionally, CE approaches can help preserve the energy embedded in the material or product, which is not feasible for any other strategies (Kirchherr et al., 2017). The process of energy recovery by incineration and landfill should be used lastly when all CE approaches have failed to work. Innovative business models are generated by the concept of CE approaches that work beyond product conservation.

An extant literature survey of CE shows that the concept has been broadly explored and its application has been analyzed through various case studies. However, research related to assessing CE's performance to measure the level of circularity of product, organizations, or geographical areas is still deficient (Haas et al., 2015). In this context, to measure and analyze a system's circularity, several authors have provided various indexed methodologies and examined the indicators relating to the systems concerned. Elia et al. (2017) analyzed the environmental assessment indicators, while Geerken et al. (2019) evaluated the contribution of public policy objectives fulfilled by CE and performed analysis based on efficiency of resources, virgin materials dependence reduction, competitiveness, job creation, and greenhouse gas emission reduction. Saidani et al. (2019) identified 55 potential circular indicators from a literature survey and categorized them into ten groups. Smol et al. (2017) have concentrated their research as area-specific and identified CE indicators related to eco-innovation in the European Union. Pauliuk (2018) has put forward a set of CE indicators based on analysis of the flow of material, accounting for the cost of the material flow, accessing the life cycle of the product, and measuring CE's performance in organizations. Five significant areas for governing CE practice rules are identified by the European Environmental Agency: input material, manufacturing, ecological design, consumption and reuse of waste, which is based on the viewpoint of the life cycle of a product.

Ghisellini et al. (2016) reviewed around 155 articles and found that only ten focused on identifying and designing the CE indicators to assess its performance, highlighting a significant gap in CE research. It is essential to understand that a lack of knowledge of CE indicators concerning academic and scientific knowledge hinders CE philosophies' application in real time (Geerken et al., 2019). Furthermore, identification of CE performance measures for rubber industries has been minimal, leaving a significant research gap in this field. This analysis aims to fill this gap by objectively evaluating and applying the global effectively factual implementation level of CE strategies to industry, goods, and services of a broader range of environmental assessment methodologies. This study identifies eight vital performance measure indicators for the rubber industry:

- **M1.** *Efficient use of resources* – optimal usage of resources like natural, human-made (polymer), workforce, machinery, transportation, etc. to balance the environment (Fan et al., 2019; Pinjing et al).
- **M2.** *Waste elimination/minimization/management* – employing various scientific methodologies and techniques to reuse rubber waste to create a usable product which can minimize waste generation and also improve

waste management (Aceleanu et al., 2019; Fan et al., 2019; Rashid et al., 2008).

M3. *Product up-grading* – up-grading the product in terms of design, material used, and enhancing the reusable capability of material (Preston, 2012; Werning and Spinler, 2020).

M4. *Product value retention* – the product's ability to be used again, providing a higher return at the end of its life cycle (Tyagi et al., 2015c; Werning and Spinler, 2020).

M5. *Eco-friendly product development* emphasizes using greener material that is environmentally friendly and does not create hazards for the ecosystem (den Hollander et al., 2017; Werning and Spinler, 2020).

M6. *Worker training and development* – it is necessary to implement CE paradigm training for the workforce to maximize CE philosophies' effectiveness (Preston, 2012; Tyagi et al., 2015c).

M7. *Managing the end of life of a product* – the end use product/material can be either fully or partially utilized in a new product (Tam et al., 2019; Toffel, 2004).

M8. *A lifetime extension of product* – utilizing modern technologies and research and development facilities increases the durability of the product to enhance its value at the end of its life (Akanbi et al., 2018; Tam et al., 2019).

All the above-mentioned performance measures provide an overview to measure the performance of CE in the rubber industry. To measure the performance of anything, some criteria should be defined based on which performance can be analyzed. For this purpose, after interaction with academics and concerned industry personnel, five criteria have been identified:

C1. *Resource protection* – CE aims to reduce the impact caused by linear economic models, which are the leading cause of destruction of the natural ecosystem. Optimized natural resource usage is needed in the current scenario with the efficient use of water, energy, and raw materials (Aceleanu et al., 2019; Klemeš et al., 2020).

C2. *Safeguarding the environment* – environment safety is of utmost importance for humanity and also for future generations. Reduction of emission level and urging use of natural resources for energy production are required in the present scenario to protect the environment from hazards (Aceleanu et al., 2019; Bhatia et al., 2020).

C3. *Customer benefit and satisfaction* – the customer is the critical element and the entire supply chain's driver. Achieving customer demand is a vital target for any industry. In today's era customers are heading towards a safer environment and waste elimination from the planet. To satisfy the customer's needs, industries should have to do the required job by implementing the concept of CE (Toffel, 2004; Tyagi et al., 2014).

C4. *Overall quality improvement* – to meet CE's requirement, the industry has to improve its product quality and durability in addition to its remanufacturing capabilities. Moreover, to make an impression in the customer's mind, industries have to change their working strategies and enable maximum usage of recyclable materials. This will improve the overall quality of the industries (de Oliveira et al., 2019; Tyagi et al., 2015d).

C5. *The financial benefit to the industry* – if an industry fulfills the customer's demands, then surely its sales will increase, which results in an increase in profit of the industry and achieves goodwill in the industrial market (Klemeš et al., 2020; Toffel, 2004).

10.3 METHODOLOGY

To complete the stated objective of this study, multi-criteria decision making has been employed. A hybrid approach of analytic hierarchy process (AHP) and elimination and choice expressing reality (ELECTRE) methods has been carried out to obtain the results. Data were collected using a questionnaire-based survey through hard copies, and then the analyzed using the above-mentioned mathematical tools.

In 1970 Bernard Roy developed a mathematical tool ELECTRE, mainly aiming to create relationships among the parameters and perform outranking relations, in order to present preferences to the decision-makers (Sawadogo and Anciaux, 2011). The ELECTRE application consists of two main components: first establishing one or more relationships aimed at comparing each of the acts in an integral way; second, a process of manipulation which takes up the recommendations obtained in the first phase. The essence of recommendation depends on the problem: collection, ranking, or sorting.

ELECTRE approach requires weighting of the criteria which has been calculated using the AHP method. Relative measurement scales are developed by the AHP. Standardization requires the dimensions of several objects to be transformed into relative measurements (Saaty, 1990). The methodology applied in this paper is elaborated step by step below:

Step-1: Assigning criteria weights through AHP: In this step, weights are obtained for the criteria using the AHP process shown in Table 10.1. These weights are used in the initial stage of the ELECTRE method. The steps of the AHP methods are described below:

(a) Construct a pairwise comparison matrix using a scale of relative importance. The judgments are entered using a scale of 1–9 as proposed by Saaty (1990), where 1 refers to equally preferred and 9 refers to i is extremely more preferred than j.

(b) Assuming p criteria, the pairwise comparison of criterion i with criterion j gives a square matrix $A_{p \times p}$, where a_{ij} denotes the relative importance of criterion i with respect to criterion j.

(c) Find the relative normalized weight (w_i) of each criterion by calculating the geometric mean of ith row and normalizing the geometric mean of rows in the comparison matrix.

TABLE 10.1
Criteria weights obtained from the analytic hierarchy process (AHP) method

	C1	C2	C3	C4	C5	Priority
C1	1	1/5	1/3	1/5	1/7	0.0436
C2	5	1	3	1	1/3	0.2017
C3	3	1/3	1	1/3	1/5	0.0888
C4	5	1	3	1	1/3	0.2017
C5	7	3	5	3	1	0.4641
λ_{max} = 5.13		CI = 0.03		CR = 3%		

CI, consistency index; CR, consistency ratio.

TABLE 10.2
Random index (RI) table for *N* criterion (Saaty, 1990)

N	1	2	3	4	5	6	7	8
RI	0.00	0.00	0.58	0.90	1.12	1.24	1.32	1.41

(d) Find out the maximum Eigenvalue which is the average of matrix A.

$$\beta_{max} = \sum_{i=1}^{n} W_i \; for\, all\; i = 1,\ldots,n \tag{1}$$

(e) Calculate the consistency index (CI):

$$CI = \left(\beta_{max} - m\right) / \left(m - 1\right) \tag{2}$$

The smaller the value of CI, the smaller is the deviation from the consistency.
(f) Calculate the consistency ratio (CR):

$$CR = CI \,/\, RI \tag{3}$$

where RI = random index for the number of criteria used in decision making obtained from Saaty's RI. for N criterion (Table 10.2) (Saaty, 1990) as follows. The result output of criteria weight is shown in Table 10.1.

Step 2: paired comparison between the stated problem's attributes and criteria has been obtained in the decision table. This assists in developing the hierarchical structure of the stated problem. An example of such a table has been presented by expert 1 in Table 10.3.

TABLE 10.3
Paired comparison between attributes and criteria obtained from expert 1

	C1	C2	C3	C4	C5
M1	9	8	4	7	2
M2	8	8	7	5	7
M3	4	6	9	9	9
M4	9	4	9	8	8
M5	9	9	6	7	6
M6	3	3	4	9	4
M7	8	8	8	8	4
M8	9	7	9	7	5

TABLE 10.4
Normalized decision matrix

	C1	C2	C3	C4	C5
M1	0.41146	0.38744	0.2325	0.32781	0.23158
M2	0.41146	0.35375	0.33214	0.26537	0.38597
M3	0.1899	0.33691	0.39856	0.39024	0.46317
M4	0.39563	0.25268	0.44838	0.34341	0.42457
M5	0.37981	0.42114	0.29892	0.34341	0.38597
M6	0.1899	0.20215	0.2325	0.40585	0.27018
M7	0.37981	0.43798	0.38196	0.37463	0.27018
M8	0.37981	0.3706	0.43178	0.35902	0.32808

Step 3: Computing normalized decision matrix: Using the below-mentioned formula, the decision matrix has been normalized and is shown in Table 10.4.

$$x_{ij} = \frac{r_{ij}}{\sqrt{\sum_{i=1}^{n} r_{ij}^2}} \qquad i = 1,2........m; \quad j = 1,2,.......n \tag{4}$$

Step 4: Computing weighted normalized decision matrix: Weights obtained from the AHP have been used to find the weighted normalized matrix presented in Table 10.5. The formula used to obtain the same has been shown below:

$$V_{ij} = w_j.x_{ij} \tag{5}$$

TABLE 10.5
Weighted normalized decision matrix

W_j	0.04361	0.20173	0.08879	0.20173	0.46414
	C1	C2	C3	C4	C5
M1	0.01794	0.07816	0.02064	0.06613	0.10749
M2	0.01794	0.07136	0.02949	0.05353	0.17915
M3	0.00828	0.06797	0.03539	0.07872	0.21498
M4	0.01725	0.05097	0.03981	0.06928	0.19706
M5	0.01656	0.08496	0.02654	0.06928	0.17915
M6	0.00828	0.04078	0.02064	0.08187	0.1254
M7	0.01656	0.08835	0.03391	0.07558	0.1254
M8	0.01656	0.07476	0.03834	0.07243	0.15227

Step 5: Calculating the concordance set: On grouping the measures, it is required to differentiate between useful and non-useful parameters. The highest and lowest values are desired for every useful and non-useful parameter, respectively. When the alternative is superior or equivalent to the other elements of the pair, it is considered under the concordance set denoted by *C*. For the function f (b_1), the alternative score is demonstrated by b_1, while w_j indicates the weight of attribute *j*. Thus the concordance index $C(b_1, b_2)$ can be demonstrated as:

$$c(b_1,b_2) = \{j,\ K_{pj} \geq K_{qj}\} \tag{6}$$

where K_{pj} is weighted evaluation of alternative A_p to the jth attribute. Or it can be stated as, $C(b_1,b_2)$ is the assembly of attributes where A_p is better than or equal to A_q. The counterpart of $C(b_1,b_2)$, which is called the discordance set that contains all attributes for which A_p is inferior than A_q, can be constructed using this formula:

$$d(b_1,b_2) = \{j,\ K_{qj} > K_{pj}\} \tag{7}$$

Step 6: Constructing concordance and discordance interval matrix: The concordance and discordance interval matrix is determined using the formula in eq. 8 and 9:

$$C(b_1,b_2) = \sum_{j=1}^{n} w_j \times c_j(b_1,b_2) \tag{8}$$

$$D(b_1,b_2) = \sum_{j=1}^{n} w_j \times d_j(b_1,b_2) \tag{9}$$

TABLE 10.6
Concordance interval matrix

	M1	M2	M3	M4	M5	M6	M7	M8	Sum
M1	0.00	0.45	0.25	0.25	0.04	0.33	0.04	0.25	1.60
M2	0.60	0.00	0.25	0.25	0.60	0.80	0.51	0.51	3.50
M3	0.75	0.75	0.00	0.87	0.75	0.80	0.75	0.67	5.35
M4	0.75	0.75	0.13	0.00	0.80	0.80	0.60	0.60	4.43
M5	0.96	0.87	0.25	0.40	0.00	0.80	0.51	0.71	4.49
M6	0.75	0.20	0.25	0.20	0.20	0.00	0.67	0.20	2.47
M7	0.96	0.49	0.25	0.40	0.54	0.80	0.00	0.45	3.88
M8	0.75	0.49	0.33	0.40	0.33	0.80	0.60	0.00	3.71
Sum	5.53	4.01	1.69	2.77	3.26	5.12	3.67	3.37	29.44

TABLE 10.7
Discordance interval matrix

	M1	M2	M3	M4	M5	M6	M7	M8	Sum
M1	0.00	1.00	1.00	1.00	1.00	0.48	1.00	1.00	6.48
M2	0.18	0.00	1.00	0.88	1.00	0.53	0.41	0.70	4.69
M3	0.09	0.27	0.00	0.50	0.47	0.04	0.23	0.13	1.73
M4	0.30	1.00	1.00	0.00	1.00	0.18	0.52	0.53	4.53
M5	0.02	0.19	1.00	0.53	0.00	0.23	0.14	0.44	2.54
M6	1.00	1.00	1.00	1.00	1.00	0.00	1.00	1.00	7.00
M7	0.08	1.00	1.00	1.00	1.00	0.13	0.00	1.00	5.21
M8	0.08	1.00	1.00	1.00	1.00	0.28	0.51	0.00	4.86
sum	1.75	5.46	7.00	5.91	6.47	1.86	3.80	4.81	37.05

The concordance and discordance interval matrix is shown in Tables 10.6 and 10.7 respectively.

Step 6: Determining concordance and discordance index: The index value of concordance and discordance set is determined by the relation in eq. 10 and 11 respectively;

$$C_{bar} = \sum_{b_1=1}^{n}\sum_{b_2=1}^{n} \frac{C_{ab}}{n(n-1)} \tag{10}$$

$$D_{bar} = \sum_{b_1=1}^{n}\sum_{b_2=1}^{n} \frac{D_{ab}}{n(n-1)} \tag{11}$$

Step 7: Determining concordance and discordance set matrix: The index matrix is constructed using the formula in eq. 12 and 13:

$$C_j(b_1,b_2)=\left\{\begin{bmatrix} 1, & if\ f_j(b_1)+q_j \geq f_j(b_2) \\ 0, & if\ f_j(b_1)+q_j \leq f_j(b_2) \\ \frac{f_j(b_1)+p_j-f_j(b_2)}{p_j-q_j} & if\ else \end{bmatrix}\right\} \quad (12)$$

$$D_{pq}=\frac{\left[\sum_{j^0}\left|v_{pj^o}-v_{qj^o}\right|\right]}{\left[\sum_{j}\left|v_{pj}-v_{qj}\right|\right]} \quad (13)$$

The concordance and discordance index matrix is shown in Tables 10.8 and 10.9.

Step 7: Calculating net concordance and discordance matrix: To find the ranking among the alternatives net a concordance and discordance matrix

TABLE 10.8
Concordance index matrix

	M1	M2	M3	M4	M5	M6	M7	M8
M1	0	0	0	0	0	0	0	0
M2	1	0	0	0	1	1	0	0
M3	1	1	0	1	1	1	1	1
M4	1	1	0	1	1	1	1	1
M5	1	1	0	0	0	1	0	1
M6	1	0	0	0	0	0	1	0
M7	1	0	0	0	1	1	0	0
M8	1	0	0	0	0	1	1	0

TABLE 10.9
Discordance index matrix

	M1	M2	M3	M4	M5	M6	M7	M8
M1	0	1	1	1	1	0	1	1
M2	0	0	1	1	1	0	0	1
M3	0	0	0	0	0	0	0	0
M4	0	1	1	0	1	0	0	0
M5	0	0	1	0	0	0	0	0
M6	1	1	1	1	1	0	1	1
M7	0	1	1	1	1	0	0	1
M8	0	1	1	1	1	0	0	0

TABLE 10.10
Ranking the performance measures using net superior and inferior values

	Net superior value	Net inferior value	Rank
M1	-3.923558532	4.732939271	8
M2	-0.512734008	-0.761949036	6
M3	3.657190801	-5.265683924	1
M4	1.660920834	-1.374462557	2
M5	1.223506406	-3.929980436	3
M6	-2.650930932	5.137746251	7
M7	0.205941135	1.406960901	5
M8	0.339664295	0.054429531	4

is needed. This helps in giving the final rank to all the alternatives. It is obtained using the formula given in eq. 14 and 15:

$$C_p = \sum_{\substack{k=1 \\ k \neq p}}^{m} C_{pk} - \sum_{\substack{k=1 \\ k \neq p}}^{m} C_{kp} \tag{14}$$

$$D_p = \sum_{\substack{k=1 \\ k \neq p}}^{m} D_{pk} - \sum_{\substack{k=1 \\ k \neq p}}^{m} D_{kp} \tag{15}$$

Using the above relations, net superior and net inferior value of the performance measures have been identified. Accordingly, based on these identified values the rank of the performance measures has been obtained, as shown in Table 10.10.

The obtained net inferior and net superior values from the ELECTRE method have been visualized in Figure 10.1.

In the figure, the net superior value is in blue and the net inferior value is represented in orange. The measures attained a positive value trend in the right direction. Here M6 has attained the highest net inferior value and M3 has gained a maximum net superior value. Moreover, M3 has achieved a least net inferior value and M1 has attained the minimum net superior value.

10.4 RESULTS AND DISCUSSION

In the present research, eight significant performance measures of CE have been acknowledged and here analyzed based on five suitable identified criteria. ELECTRE method was used to provide an outranking of the selected parameters and additionally to provide interrelationship between them. ELECTRE method requires weighting of the criteria, and AHP was used in this research to obtain a consistency ratio of 0.03. Weights obtained from AHP have been directly used in the ELECTRE method. The outcome shows that the performance measure

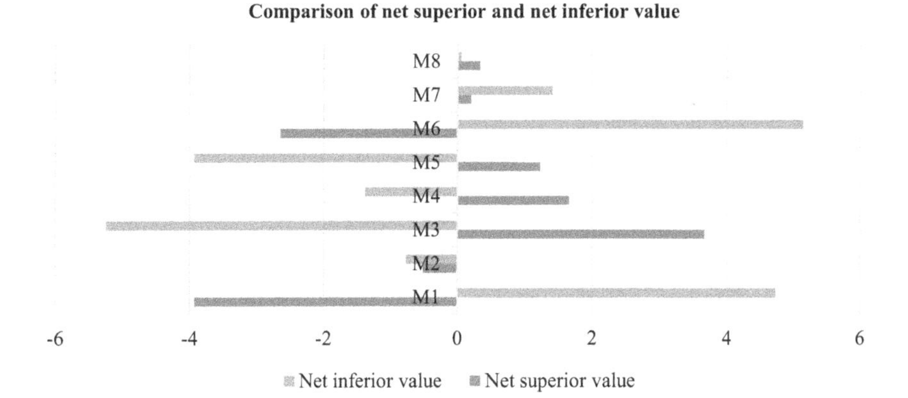

FIGURE 10.1 Representation of net inferior and net superior value of performance measures.

M3 (product up-grading) has obtained the top rank among all the parameters. Up-grading of the product signifies an enhancement of the product features, using more durable product material which can enable its reusability and sustain its use for longer. This can be implemented by making use of the research and development facilities within the organization and with back-end demand from the customer (Klemeš et al., 2020). Government rules and regulations will also contribute to the same purpose (Moktadir et al., 2018). The second most crucial performance found as an outcome in this research is M4 (product value retention) which signifies the maximum usage of a product or its containing material, which is also an outcome of product up-grading. In the current research, the order of performance measure is obtained as M3>M4>M5>M8>M7>M2>M6>M1. As depicted from the results M1 (efficient use of resources) is found to be lower-ranked, but its impact on all of the other measures is very high.

The results from the present research show that the significant need of society in the present and near future is healthy living, which can be solved by producing an eco-friendly product (M5). Managing the end of life of the product (M7) and increasing the lifetime of the product (M8) will assist in increasing the usability of the product and thus decrease the exploitation of natural resources, resulting in their conservation for future generations, which in turn helps in waste minimization (M2). Implementation of all of the concepts of CE training of the workforce (M6) is of utmost importance to achieve the required goals.

The ranking of CE's performance measures depicts an overall clarity for policymakers and managers, which will enable a check for their working environment. As an outcome of the present research, the prioritized performance measures aid the enterprise managers and top leaders to check the following CE practices and subsequently enhance CE viewpoints. The prioritizations assist in developing strategies to measure the performance of implied CE principles and thus govern better CE practices throughout. In the case of rubber industries there are hinerances in the recycling and reusing process due to hybrid material usage.

The CE process could be enhanced by making the product more durable and enhancing its life cycle, contributing to overall resource conservation in the long run and minimizing waste, aligning with the CE philosophies. The result of the present research will enable a checklist to be drawn up for managers to map their actual practices.

10.5 CONCLUSION AND FUTURE SCOPE

The performance measures suggest the effectiveness of CE philosophies which successfully increase the sustainability practices of the industries. Performance measures analyze CE's performance in any industry – in this study, the rubber industry. Rubber industries were chosen in this research because the by-products and product end waste from this industry are very hazardous for society and the environment. As the need for environmental protection is top of the agenda these days (Genovese et al., 2017), all industries, whether small or medium enterprises or large enterprises, are aware of this theme and are trying to incline their processes towards environment protection (Mura et al., 2020). Rubber industries are also heading in the same direction and are aware of resource protection and environmental safety. However, due to some hindrances, they are not adequately involved (Svoboda et al., 2018). The aim of this research was to generate awareness about the concept of CE in\policymakers. The present study's findings will enable managers to build an effective and reliable CE supply chain network by enhancing a company's overall productivity and maintaining the safety of the environment.

The proposed study establishes a path to boost CE practices. Besides contributing towards mitigating environmental problems, this study also focuses on the financial benefit for industries of using CE philosophies. As the awareness of CE and pressure from governments towards sustainable practices grow, this aids institutions in implementing practices which would contribute to environmental safety. But to check the effectiveness of the operations performed is a probabilistic task for industries and this research contributes to the same by proposing a framework with the capability of measuring CE performance. Overall, in a nut shell, it can be said that the current research aims to propose a framework that signifies the effectiveness of CE.

Although this research has wide application in rubber industry products in automobile sectors, its scope is limited. In this research, specified rubber industries (automobile-based) were chosen and these cannot be generalized to all rubber industries manufacturing rubber-made products. Respondents were also from a specific area (Punjab), so the present research cannot be generalized all over the country, as different zones have different limitations and advantages. Secondly, to validate current research, other multi-criteria decision-making tools could be used. In addition, case studies can be taken into account to measure the performance of an industry that is already working on the principle of CE. As a result, areas where the industry is lacking may be discovered, providing a generalized working procedure for other similar types of industries.

REFERENCES

Aceleanu, M.I., Serban, A.C., Suciu, M.C., Bitoiu, T.I., 2019. The management of municipal waste through circular economy in the context of smart cities development. IEEE Access 7, 133602–133614. https://doi.org/10.1109/ACCESS.2019.2928999

Akanbi, L.A., Oyedele, L.O., Akinade, O.O., Ajayi, A.O., Davila Delgado, M., Bilal, M., Bello, S.A., 2018. Salvaging building materials in a circular economy: A BIM-based whole-life performance estimator. Resour. Conserv. Recycl. 129, 175–186. https://doi.org/10.1016/j.resconrec.2017.10.026

Bhatia, M.S., Jakhar, S.K., Mangla, S.K., Gangwani, K.K., 2020. Critical factors to environment management in a closed loop supply chain. J. Clean. Prod. 255. https://doi.org/10.1016/j.jclepro.2020.120239

den Hollander, M.C., Bakker, C.A., Hultink, E.J., 2017. Product design in a circular economy: Development of a typology of key concepts and terms. J. Ind. Ecol. 21, 517–525. https://doi.org/10.1111/jiec.12610

de Oliveira, C.T., Mônica, M.M.M., Campos, L.M.S., 2019. Understanding the Brazilian expanded polystyrene supply chain and its reverse logistics towards circular economy. J. Clean. Prod. 235, 562–573. https://doi.org/10.1016/j.jclepro.2019.06.319

Elia, V., Gnoni, M.G., Tornese, F., 2017. Measuring circular economy strategies through index methods: A critical analysis. J. Clean. Prod. 142, 2741–2751. https://doi.org/10.1016/j.jclepro.2016.10.196

Esposito, M., Tse, T., Soufani, K., 2018. Introducing a circular economy: New thinking with new managerial and policy implications. Calif. Manage. Rev. 60, 5–19. https://doi.org/10.1177/0008125618764691

Ethirajan, M., Kandasamy, J., 2019. An analysis on sustainable supply chain for circular economy. Procedia Manufacturing. Elsevier, pp. 477–484. https://doi.org/10.1016/j.promfg.2019.04.059

Fan, Y.V., Lee, C.T., Lim, J.S., Klemeš, J.J., Le, P.T.K., 2019. Cross-disciplinary approaches towards smart, resilient and sustainable circular economy. J. Clean. Prod. https://doi.org/10.1016/j.jclepro.2019.05.266

Farooque, M., Zhang, A., Thürer, M., Qu, T., Huisingh, D., 2019. Circular supply chain management: A definition and structured literature review. J. Clean. Prod. https://doi.org/10.1016/j.jclepro.2019.04.303

Geerken, T., Schmidt, J., Boonen, K., Christis, M., Merciai, S., 2019. Assessment of the potential of a circular economy in open economies – Case of Belgium. J. Clean. Prod. 227, 683–699. https://doi.org/10.1016/j.jclepro.2019.04.120

Genovese, A., Acquaye, A.A., Figueroa, A., Koh, S.C.L., 2017. Sustainable supply chain management and the transition towards a circular economy: Evidence and some applications. Omega (United Kingdom) 66, 344–357. https://doi.org/10.1016/j.omega.2015.05.015

Ghisellini, P., Cialani, C., Ulgiati, S., 2016. A review on circular economy: The expected transition to a balanced interplay of environmental and economic systems. J. Clean. Prod. 114, 11–32. https://doi.org/10.1016/j.jclepro.2015.09.007

Haas, W., Krausmann, F., Wiedenhofer, D., Heinz, M., 2015. How circular is the global economy? An assessment of material flows, waste production, and recycling in the European union and the world in 2005. J. Ind. Ecol. 19, 765–777. https://doi.org/10.1111/jiec.12244

Kirchherr, J., Reike, D., Hekkert, M., 2017. Conceptualizing the circular economy: An analysis of 114 definitions. Resour. Conserv. Recycl. 127, 221–232. https://doi.org/10.1016/j.resconrec.2017.09.005

Klemeš, J.J., Fan, Y.V., Tan, R.R., Jiang, P., 2020. Minimising the present and future plastic waste, energy and environmental footprints related to COVID-19. Renew. Sustain. Energy Rev. 127. https://doi.org/10.1016/j.rser.2020.109883

Mangla, S.K., Luthra, S., Mishra, N., Singh, A., Rana, N.P., Dora, M., Dwivedi, Y., 2018. Barriers to effective circular supply chain management in a developing country context. Prod. Plan. Control 29, 551–569. https://doi.org/10.1080/09537287.2018.1449265

Moktadir, M.A., Rahman, T., Rahman, M.H., Ali, S.M., Paul, S.K., 2018. Drivers to sustainable manufacturing practices and circular economy: A perspective of leather industries in Bangladesh. J. Clean. Prod. 174, 1366–1380. https://doi.org/10.1016/j.jclepro.2017.11.063

Morseletto, P., 2020. Targets for a circular economy. Resour. Conserv. Recycl. 153, 104553. https://doi.org/10.1016/j.resconrec.2019.104553

Mura, M., Longo, M., Zanni, S., 2020. Circular economy in Italian SMEs: A multi-method study. J. Clean. Prod. 245, 118821. https://doi.org/10.1016/j.jclepro.2019.118821

Murray, A., Skene, K., Haynes, K., 2017. The circular economy: An interdisciplinary exploration of the concept and application in a global context. J. Bus. Ethics 140, 369–380. https://doi.org/10.1007/s10551-015-2693-2

Pauliuk, S., 2018. Critical appraisal of the circular economy standard BS 8001:2017 and a dashboard of quantitative system indicators for its implementation in organizations. Resour. Conserv. Recycl. 129, 81–92. https://doi.org/10.1016/j.resconrec.2017.10.019

Preston, F., 2012. A global redesign? Shaping the circular economy. Energy, Environ. Resour. Gov. 1–20. https://doi.org/10.1080/0034676042000253936

Rashid, S.H.A., Evans, S., Longhurst, P., 2008. A comparison of four sustainable manufacturing strategies. Int. J. Sustain. Eng. https://doi.org/10.1080/19397030802513836

Saaty, T.L., 1990. How to make a decision: The analytic hierarchy process. Eur. J. Oper. Res. 48, 9–26. https://doi.org/10.1016/0377-2217(90)90057-I

Saidani, M., Yannou, B., Leroy, Y., Cluzel, F., Kendall, A., 2019. A taxonomy of circular economy indicators. J. Clean. Prod. https://doi.org/10.1016/j.jclepro.2018.10.014

Sawadogo, M., Anciaux, D., 2011. Intermodal transportation within the green supply chain: An approach based on ELECTRE method. Int. J. Bus. Perform. Supply Chain Model. 3, 43–65. https://doi.org/10.1504/IJBPSCM.2011.039973

Smol, M., Kulczycka, J., Avdiushchenko, A., 2017. Circular economy indicators in relation to eco-innovation in European regions. Clean Technol. Environ. Policy. https://doi.org/10.1007/s10098-016-1323-8

Suhi, S.A., Enayet, R., Haque, T., Ali, S.M., Moktadir, M.A., Paul, S.K., 2019. Environmental sustainability assessment in supply chain: An emerging economy context. Environ. Impact Assess. Rev. 79, 106306. https://doi.org/10.1016/j.eiar.2019.106306

Svoboda, J., Vaclavik, V., Dvorsky, T., Klus, L., Zajac, R., 2018. The potential utilization of the rubber material after waste tire recycling. IOP Conf. Ser. Mater. Sci. Eng. 385. https://doi.org/10.1088/1757-899X/385/1/012057

Tam, E., Soulliere, K., Sawyer-Beaulieu, S., 2019. Managing complex products to support the circular economy. Resour. Conserv. Recycl. 145, 124–125. https://doi.org/10.1016/j.resconrec.2018.12.030

Toffel, M.W., 2004. Strategic management of product recovery. Calif. Manage. Rev. https://doi.org/10.2307/41166214

Tyagi, M., Kumar, P., Kumar, D., 2014. A hybrid approach using AHP-TOPSIS for analyzing e-SCM performance. Procedia Engineering. Elsevier, pp. 2195–2203. https://doi.org/10.1016/j.proeng.2014.12.463

Tyagi, M., Kumar, P., Kumar, D., 2015a. Analyzing CSR issues for supply chain performance system using preference rating approach. J. Manuf. Technol. Manag. 26, 830–852. https://doi.org/10.1108/JMTM-03-2014-0031

Tyagi, M., Kumar, P., Kumar, D., 2015b. Parametric selection of alternatives to improve performance of green supply chain management system. Procedia - Soc. Behav. Sci. 189, 449–457. https://doi.org/10.1016/j.sbspro.2015.03.197

Tyagi, M., Kumar, P., Kumar, D., 2015c. Assessment of critical enablers for flexible supply chain performance measurement system using fuzzy DEMATEL approach. Glob. J. Flex. Syst. Manag. 16, 115–132. https://doi.org/10.1007/s40171-014-0085-6

Tyagi, M., Kumar, P., Kumar, D., 2015d. Analysis of interactions among the drivers of green supply chain management. Int. J. Bus. Perform. Supply Chain Model. 7, 92–108. https://doi.org/10.1504/IJBPSCM.2015.068137

Werning, J.P., Spinler, S., 2020. Transition to circular economy on firm level: Barrier identification and prioritization along the value chain. J. Clean. Prod. 245, 118609. https://doi.org/10.1016/j.jclepro.2019.118609

11 Fuzzy FMEA Application in the Healthcare Industry

Prateek Saxena, Dilbagh Panchal, and Mohit Tyagi
Department of Industrial and Production Engineering, Dr. B. R. Ambedkar National Institute of Technology Jalandhar, Punjab, India

11.1 INTRODUCTION

With the rise of accidental injury in India, trauma care has become a major cause of concern. According to the 2018 National Crime Records Bureau (NCRB) report on India (https://ncrb.gov.in/accidental-deaths-suicides-india-2018), the rate of accidental deaths (per lakh of population) has increased from 30.3 in 2017 to 31.1 in 2018. Trauma injuries, which are among the leading causes of mortality, pose a greater challenge to physicians. These life-threatening injuries are some of the most difficult decisions physicians face. Due to inadequate resources, district hospitals are unable to manage trauma patients, which may lead to delay in treatment and sometimes results in poor outcome. Hospitals lacking a sufficient emergency centre cannot handle patients with the most extreme injuries. The mortality rate can be decreased through proper modifications in hospital administration and training. Recently, training in trauma life support has become available. To remove inadequacy in trauma systems some initiatives have been seen recently. In order to reduce risk in a trauma management system, proper systematic evaluation should be done. To handle such issues the failure mode and effect analysis (FMEA) decision method is of supreme importance. This approach allows the analyser to summarize complete qualitative information; this is useful for the accuracy of the results required to be implemented to overcome bottlenecks faced by the hospital in imparting quality services to their patients. With some advantages, the FMEA approach also has limitations which many researchers have noted (Sharma and Sharma, 2012; Panchal and Kumar, 2016).

To overcome these limitations, Wang et al. in 2009 developed a fuzzy methodology-based FMEA approach in which O, S and D are fuzzified with suitable membership function and using fuzzy logic a rule-based inference model was built. Fuzzy logic has the following advantages over classical FMEA:

DOI: 10.1201/9781003181644-11

- With the use of linguistic terms, risk can be evaluated directly together with potential failure.
- Vagueness in the assessment of the system can be eliminated.
- It is able to model complex non-linear functions quickly.

In the past, many authors have applied fuzzy FMEA to safety and risk analysis. To name a few, Sutrisnoa et al. (2015) applied a modified FMEA model to access the risk of maintenance waste. Chanamool and Naenna (2016) expounded the application of fuzzy FMEA in a decision-making problem for the emergency department. Dag˘suyu et al. (2016) presented a fuzzy FMEA tool-based application for studying risk issues in a sterilization unit. Renjith et al. (2018) applied fuzzy failure mode effect and criticality analysis (FMECA) of liquefied natural gas storage facility. Mutlu and Altuntas (2019) proposed a new integrated FMEA, fault tree analysis (FTA) and belief in fuzzy probability estimations of time (BIPFET) approach-based model for risk analysis for occupational safety and health in the textile industry. Xu et al. (2019) implemented an FMEA tool for evaluating radiotherapy treatment delay. Mardani et al. (2019) developed a decision-making tool based on fuzzy sets theory to evaluate healthcare and medical problems. From this review of the literature it is clear that the FMEA applied in uncertain conditions has not been implemented by any researcher to improve the service quality of a trauma centre hospital in the Indian context. Considering this as a gap in the research, the authors have applied the fuzzy rule base model to improve the service quality of the trauma emergency centre.

11.2 NOTION OF FUZZY SET

11.2.1 Fuzzy and Crisp Numbers

A crisp set can be defined by membership function $\mu_A(x)$, as shown in Eq. (1):

$$\mu_A(x) = \begin{cases} 1, & if\ x \in A \\ 0, & if\ x \in A \end{cases} \tag{1}$$

Hence, crisp set can either be 0 or 1.

A fuzzy set can be defined by membership function $\overline{\mu}_A(x)$, as shown in Eq. (2):

$$\overline{\mu}_A(x) = \{ (x, \mu_A(x)) \mid x \in X \} \tag{2}$$

Hence, a fuzzy set can have any real numbers between 0 and 1.

11.2.2 Membership Function

In the proposed study trapezoidal membership functions (MFs) have been considered for analysing the service quality issues of the trauma emergency system.

A trapezoidal MF (TRMF) is defined by Eq. (3) as:

$$\bar{\mu}_s(t) = \begin{Bmatrix} \frac{x-a_0}{b_0-a_0}, a_0 \le t \le b_0 \\ 1, b_0 \le t \le c_0 \\ \frac{d_0-x}{d_0-c_0}, c_0 \le t \le d_0 \\ 0, \textit{otherwise} \end{Bmatrix} \tag{3}$$

where $\bar{\mu}_S(t)$ is the TRMF with fuzzy numbers as (a_0, b_0, c_0, d_0).
a_1 and b_1 are pre and post values of b_0
a_2 and b_2 are pre and post values of c_0.

11.3 FMEA APPROACH

FMEA is one of the methods that can be used for risk assessment of any system. In this method, the potential failures and their causes are identified in the risk analysis. To begin FMEA, a system must be understood so that every possible potential failure and its effects can be analysed. The procedure of fuzzy FMEA that has been carried out is illustrated in the following steps:

Step 1. For fuzzy evaluation scales of occurrence (O), severity (S) and non-detection (D) must be defined. The scales were divided into five levels andh scored so that the membership function for O, S and D could be generated, as shown in Table 11.1. This step is necessary in order to convert input variables of O, S and D to fuzzy inputs.

Step 2. Due to extreme vagueness, the fuzzy risk (output) was considered to be TRMF and divided into nine levels.

Step 3. The rules were defined by the experts experienced in the system (trauma emergency centre). Comparing levels of these three factors, we obtained 5*5*5 = 125 rules. In this study, the MATLAB program was used to define the rules.

Step 4. To convert fuzzified fuzzy risk priority number (FRPN) values into defuzzified values, a centroid method was used in the MATLAB program. This it is represented mathematically by Eq. (4):

$$\tilde{x} = \frac{\int_{x_1}^{x_2} x.\, \mu_s(x)dx}{\int_{x_1}^{x_2} \mu_s(x)dx} \tag{4}$$

Step 5. The defuzzified output values were arranged in descending order, which signifies the priority of one failure over another.

TABLE 11.1
Rating and meaning of severity, occurrence and non-detection

Rating	Meaning	Severity	Occurrence	Non-detection
1	None	No effect on the system or patient	No observed failure (1 in 1240 cases observed)	Detected 9/10
2, 3	Low	Minor disruption (minimal effect on system or patient)	Rare failure (1 in 600 cases observed)	Detected 7/10
4, 5	Medium	Major disruption in operations (which can be reworked or avoided)	Occasional failure (1 in 300 cases observed)	Detected 5/10
6, 7, 8	High	Major disruption in operations (which cannot be reworked or avoided)	Frequent failure (1 in 100 cases observed)	Detected 3/10
9, 10	Very high	May endanger the patient without warning	Common failure (1 in 25 cases observed)	Detected 1/10

11.4 CASE STUDY

To exemplify the application of the proposed framework, a trauma emergency centre in Uttar Pradesh, India was considered in the present study. By applying fuzzy FMEA technique different failure causes occurring in the trauma emergency centre can be eliminated and it provides an application for improving the strategic decision-making process. The trauma emergency centre consists of four stages: pre-hospital phase, triage, diagnostic centre and referral centre.

11.4.1 Application of FMEA and Fuzzy FMEA

On the basis of expert opinion causes of failure that may occur in the trauma emergency centre are shown in Table 11.2.

The rating of severity, occurrence and non-detection is shown in Table 11.1.

Based on these ratings, scoring was conducted by a team of three experts, as shown in Table 11.3.

After scoring, the risk priority number (RPN) is calculated based on classical FMEA and FRPN is calculated using MATLAB software. The results are compared as shown in Table 11.4.

The fuzzy logic was implemented using MATLAB R2014b software, as shown in Figure 11.1.

Membership function for input as well as output wa generated (Figure 11.2).

After defining the input (severity, occurrence, non-detection) and output (fuzzy risk), 125 if–then rule bases were generated. Fuzzy inference was used to evaluate each rule, as shown in Figure 11.3.

TABLE 11.2
Causes of failure in trauma emergency centre

S no.	Failure cause
1	Delayed response time because of miscommunication between the ambulance and hospital (C_1)
2	The paramedics or emergency medical technicians lack the experience and specialty (C_2)
3	Primary survey (ABCDE) is inaccurate because of lack of experience and specialty (C_3)
4	Confusion arises in screening the patient, stabilization and admission in one of the zones (red, yellow and green) (C_4)
5	Delay in transfer to intensive care unit because proper information is not obtained from emergency medicine team (C_5)
6	Delay in proper arrangements due to error in planning process by the hospital (C_6)
7	Mistakes in clinical details, investigation reports and reviews (C_7)
8	Waiting for prescription medications because of unavailability of medications or drugs (C_8)

TABLE 11.3
Failure mode and effect analysis (FMEA) table

	Expert 1			Expert 2			Expert 3			Average		
Failure cause	O	S	D	O	S	D	O	S	D	O_e	S_e	D_e
C_1	9	4	5	9	9	3	2	3	3	6.66	5.33	3.66
C_2	6	6	2	2	6	4	8	7	6	5.33	6.33	4.00
C_3	7	4	6	2	9	3	9	9	9	6.00	7.33	6.00
C_4	6	3	3	2	5	5	5	4	4	4.33	4.00	4.00
C_5	1	4	9	4	9	1	8	6	6	4.33	6.33	5.33
C_6	2	6	1	4	5	4	5	4	4	3.66	5.00	3.00
C_7	1	6	4	3	4	3	7	4	6	3.66	4.66	4.33
C_8	1	1	1	3	4	1	8	4	6	4.00	3.00	2.66

O = occurrence; S = severity; D = non-detection.

A surface plot showing the relation between output and input is shown in Figure 11.4.

11.5 RESULTS AND DISCUSSIONS

Table 11.4 represents the comparison of rankings of failure causes with the implementation of the FMEA and fuzzy FMEA approach. Here, cause C_3 is found to

TABLE 11.4
Result comparison between failure mode and effect analysis (FMEA) and fuzzy FMEA

Failure cause	RPN = O_e*S_e*D_e	Priority	FRPN	Ranking
C_1	130.3704	4	566	4
C_2	135.1111	3	639	3
C_3	264.0000	1	799	1
C_4	69.3300	6	483	6
C_5	143.3704	2	665	2
C_6	55.0000	7	445	7
C_7	74.14815	5	516	5
C_8	32	8	347	8

RPN = risk priority numbers; O = occurrence; S = severity; D = non-detection; FRPN = fuzzy risk priority number.

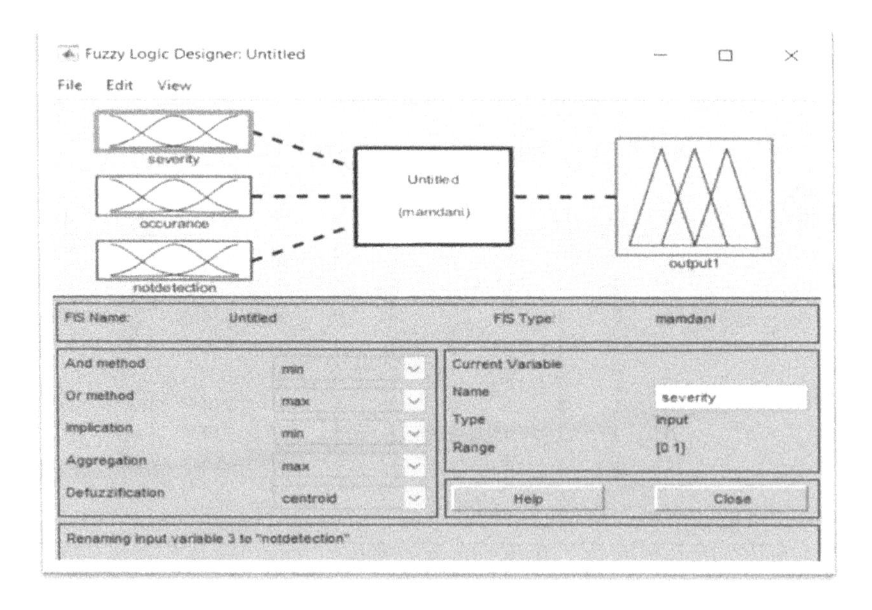

FIGURE 11.1 Fuzzy logic in MATLAB.

be the riskiest one affecting the service quality of the considered hospital. The output of FMEA and fuzzy FMEA gives the same ranking among the other listed failure causes. Special care should be taken to minimize or eliminate the effect of the causes.

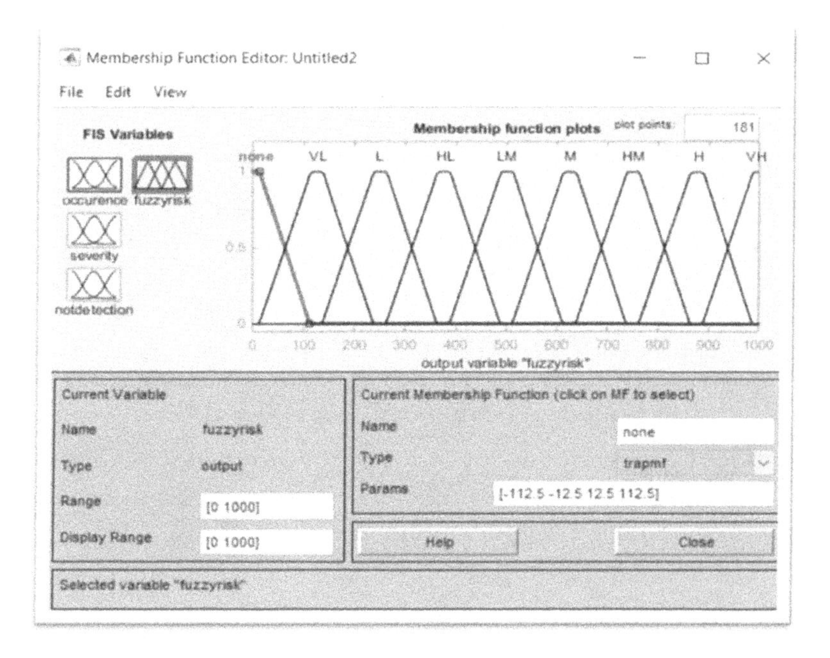

FIGURE 11.2 Membership function for output of fuzzy risk priority numbers (RPN).

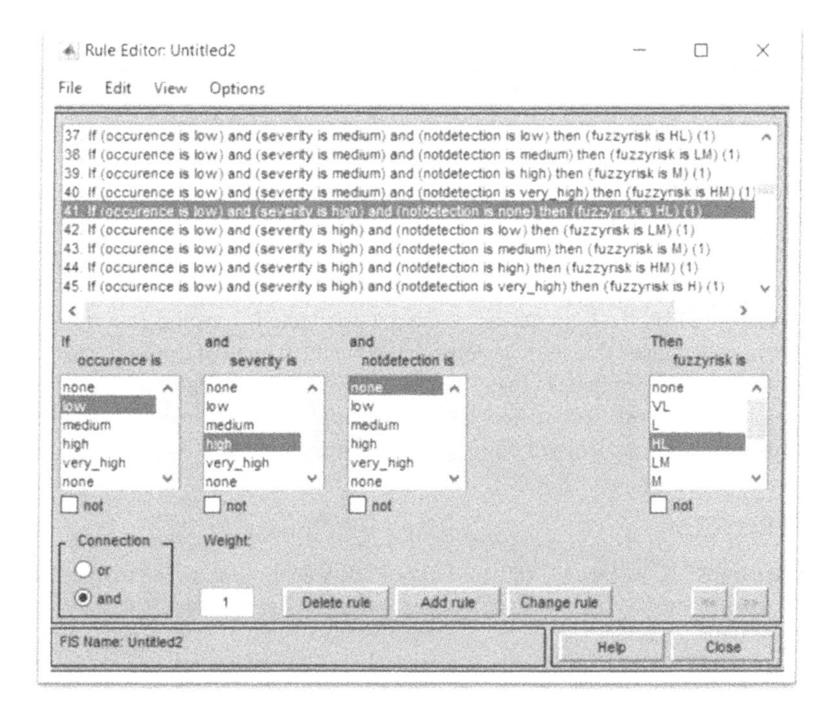

FIGURE 11.3 If–then rule base.

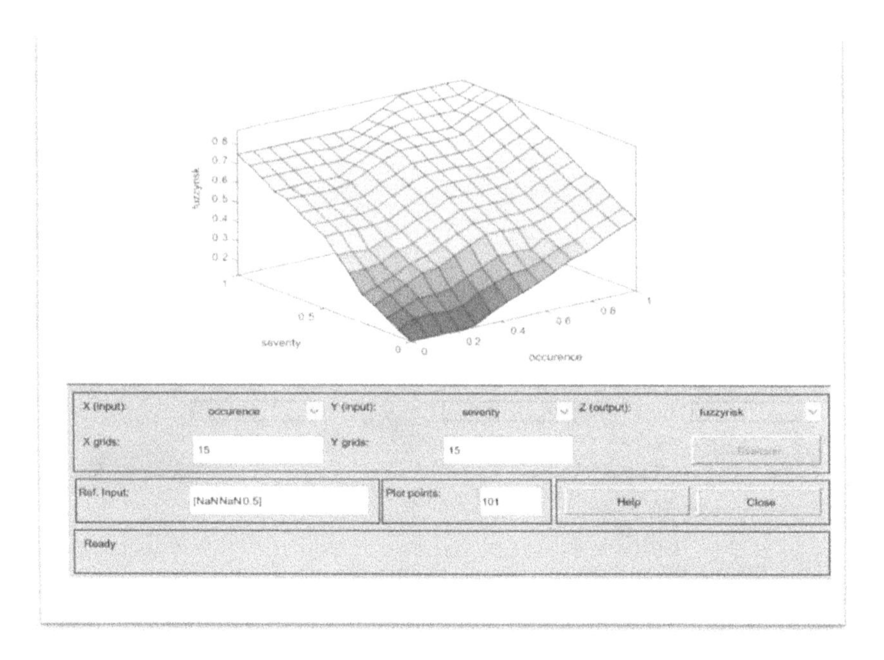

FIGURE 11.4 Surface plot between output (fuzzy risk) and input (severity, occurrence).

11.6 CONCLUSION

The risk analysis of the trauma emergency centre has been illustrated by a fuzzy FMEA approach. In the present study, the authors used an integrated fuzzy FMEA framework for the selection of a proper strategy. When considering the vagueness of the collected raw qualitative data a fuzzy concept is used. Cause C_3 was found to be the riskiest one. With an increase in the number of accidents in India, decision making has become a challenge for physicians as well as for the hospital administration and the results of this study could be useful to build confidence in the service quality of a trauma emergency centre. Further, analysis of the results is based on the correctness of the data collected from the experts. These results may be biased due to their dependency on the source of the data. However, when considering vagueness fuzzy set theory was incorporated with the FMEA approach in this work. The results have been supplied to the hospital administration for validation, which will be judged after implementation.

REFERENCES

Chanamool, N., & Naenna, T. (2016). Fuzzy FMEA application to improve decision-making process in an emergency department. *Applied Soft Computing*, *43*, 441–453.

Dağsuyu, C., Göçmen, E., Narlı, M., & Kokangül, A. (2016). Classical and fuzzy FMEA risk analysis in a sterilization unit. *Computers & Industrial Engineering*, *101*, 286–294.

Mardani, A., Hooker, R. E., Ozkul, S., Yifan, S., Nilashi, M., Sabzi, H. Z.., & Fei, G. C. (2019). Application of decision making and fuzzy sets theory to evaluate the healthcare and medical problems: a review of three decades of research with recent developments. *Expert Systems with Applications*, *137*, 202–231.

Mutlu, N. G., & Altuntas, S. (2019). Risk analysis for occupational safety and health in the textile industry: integration of FMEA, FTA, and BIFPET methods. *International Journal of Industrial Ergonomics*, *72*, 222–240.

Panchal, D., & Kumar, D. (2017). Risk analysis of compressor house unit in thermal power plant using integrated fuzzy FMEA and GRA approach. *International Journal of Industrial and Systems Engineering*, *25*(2), 228–250.

Renjith, V. R., Kumar, P. H., & Madhavan, D. (2018). Fuzzy FMECA (failure mode effect and criticality analysis) of LNG storage facility. *Journal of Loss Prevention in the Process Industries*, *56*, 537–547.

Sharma, R. K., & Sharma, P. (2012). Integrated framework to optimize RAM and cost decisions in a process plant. *Journal of Loss Prevention in the Process Industries*, *25*(6), 883–904.

Sutrisno, A., Gunawan, I., & Tangkuman, S. (2015). Modified failure mode and effect analysis (FMEA) model for accessing the risk of maintenance waste. *Procedia Manufacturing*, *4*, 23–29.

Wang, Y. M., Chin, K. S., Poon, G. K. K., & Yang, J. B. (2009). Risk evaluation in failure mode and effects analysis using fuzzy weighted geometric mean. *Expert Systems with Applications*, *36*(2), 1195–1207.

Xu, Z., Lee, S., Albani, D., Dobbins, D., Ellis, R. J., Biswas, T., ... & Podder, T. K. (2019). Evaluating radiotherapy treatment delay using failure mode and effects analysis (FMEA). *Radiotherapy and Oncology*, *137*, 102–109.

12 Drivers of Industry 4.0 in a Circular Economy Initiative in the Context of Emerging Markets

Chitranshu Khandelwal,[1] *Sourabh Kumar,*[1] *and Mukesh Kumar Barua*[2]

[1] Ph.D. Students, Department of Management Studies, Indian Institute of Technology Roorkee, Haridwar, India

[2] Professor, Department of Management Studies, Indian Institute of Technology Roorkee, Haridwar, India

12.1 INTRODUCTION

It is challenging to manage the adverse effects of unsustainable and environmental degradation patterns in the era of globalization. Sustainability requires exploration of improvements in the resource's efficiency in the global economy (Rajput & Singh, 2019). In recent years academicians, researchers, and other stakeholders hav increasingly been exploring circular economy (CE) and Industry 4.0 (I4.0) (Luiz Mattos Nascimento et al., 2019). I4.0 is supported by the advancement of data security and transparency in the supply chain process (Papadopoulos et al., 2017). It is necessary to integrate disruptive technologies in workflows to obtain continuous improvement in the process. These disruptive technologies consist of 3D printing, artificial intelligence, machine learning, augmented reality, automation, robotics, blockchain, edge computing, and so forth (Baaziz & Quoniam, 2014; Bumblauskas et al., 2019; Sousa et al., 2019). These new digital technologies help the CE bring transparency, security, and trust to the supply chain (Norta et al., 2019). There is a great need for both researchers and business communities to explore further digital technologies in the production and supply chain. Several studies have been published on the benefits of I4.0 and CE. For instance Parast et al. (2011), Xu et al. (2011), Lele (2019), and Tjahjono et al. (2017) have elaborated on the benefits of operational efficiency, tracking, and security, reduction in waste, counterfeiting, and transparency.

A CE is a systematic economic development approach to benefit business, the environment, and society (S. Kumar et al., 2021). It uses a circular model

DOI: 10.1201/9781003181644-12

rather than a linear "take–make–waste" economic model (Abdul-hamid et al., 2020). Luiz Mattos Nascimento et al. (2019) explored the integrated technology advancement business model of the CE that reuses discarded material. The business model suggests that scrap electronics devices and reverse logistics support CE practices. In the CE, multiple industries can use a single supply chain network through industrial symbiosis (Tseng et al., 2018).

The CE offers guidelines and principles for promoting sustainable development. In the CE ecosystem, components, products, and by-products do not lose their value. If products fail to meet the quality standards, CE considers the holistic perspective, from production to consumers and vice versa (Casado-Vara et al., 2018). It also improves resource allocation by strengthening the connection between industries and network structural changes (Luiz Mattos Nascimento et al., 2019).

For sustainable supply chain management, managers have to identify and implement new technologies in the production and distribution process. The existing research suggests insufficient research investigating the determinant factors of I4.0 for the CE initiative. Our research aims to explore and fill this research gap. In the present research, we used ISM and MicMac methodology to establish the drivers' dependence relationship among the determinants. The study's objectives are to bridge the literature gap by considering the various drivers of I4.0 for the CE. In this respect, there were four research objectives :

1. To determine the drivers of I4.0 for the CE initiative in the emerging market.
2. To develop the structural model by considering the enablers of I4.0 for CE in the emerging market.
3. To calculate the driving and dependence power of the consisting drivers.
4. To develop the managerial implications of the study.

The structure of the paper is as follows: a literature review of CE and I4.0 is presented in section 12.2. The research methodology of interpretive structural modeling and MicMac analysis is presented in section 12.3. The results of the study are elaborated on in section 12.4. The implications of the study are explained in section 12.5. Section 12.6, the final section, describes the conclusions of the research, including its limitations and future research directions.

12.2 LITERATURE REVIEW

In sustainable business practices, it is essential to explore further the CE drivers and I4.0. The CE transforms production and consumption by creating useful products from waste (Kouhizadeh et al., 2019). I4.0 can assist organizations with inflexible and diversified production systems. Rajput and Singh (2019) identified the connection between I4.0 and CE. Authors used principal component analysis (PCA) techniques to identify the prime factors and DEMATEL to establish cause-and-effect relationships. The findings show that artificial intelligence, operation, and policy structure are key enablers. By implementing digital technologies in the logistics supply chain, organizations can assess real-time resource allocation (Lele, 2019).

Gupta et al. (2021) formulated a CE-based framework to assess the performance of the organizations. The authors used the best–worst method to prioritize identified practices. The results suggested building sustainability in the organization's streamlined operations: top managements need to focus on implementing CE practices.

S. Kumar et al. (2021) determined the barriers in the digitalized CE. The authors considered the case of the agricultural supply chain and used a hybrid methodology, ISM-ANP (analytic network process). The results ascertained that insufficient government support is the main obstacle to implementing the CE and I4.0.

Abdul-hamid et al. (2020) identified challenges related to the digitalized CE environment. The authors used the Delphi method to screen out unimportant attributes. They used interpretive structural modeling to understand the relationships between the challenge's practices.

P. Kumar et al. (2021) analyzed the I4.0 and CE barriers to improving the sustainability of the supply chain. The authors identified nine sustainability criteria and 15 barriers in the digitalized CE structural environment. They used an integrated methodology of analytic hierarchy process (AHP) and elimination and choice expressing reality (ELECTRE) to map sustainability criteria. The results suggest that resource circularity and energy efficiency are the prime sustainability practices. By implementing the Internet of Things (IoTs) technologies, companies can carry out extensive data cleansing using advanced machine learning algorithms to generate valuable insights (Rajput & Singh, 2018).

Zhou et al. (2020) explored the energy and environmental-based driving forces of I4.0 and CE. The results demonstrated that pollution abetment technologies and backstop technologies are the primary drivers of sustainable growth.

Our study determines the drivers of I4.0 for the CE initiative through a literature review and expert panel. The expert team consisted of 25 members from a diverse field of experience in the manufacturing and logistics sectors. The expert team verified all the drivers of the system and gave two new drivers (cloud-based design and manufacturing and health and safety) for the study. A detailed list of drivers is shown in Table 12.1.

12.3 SOLUTION METHODOLOGY

In this study, interpretative structural modeling (ISM) methodology was initially explained by Warfield (1973), used in complex decision conditions as a communication tool. ISM breaks down a complex problem into subcomponents and establishes a structural model. ISM has been widely used by scholars in a variety of fields of business management to frame strategies and make decisions in recent years (Govindan et al., 2012; Diabat et al., 2014; Gopal & Thakkar, 2016; Kamble et al., 2018; Mangla et al., 2018; Khandelwal & Barua, 2020).

TABLE 12.1
List of identified drivers of Industry 4.0 (I4.0)

Drivers	Code	Description	References
Management strategy and commitment	MSC	Strategy and commitment of top management for circular initiatives in business practices lead to attaining overall sustainability goals and ethical, long-term business growth	Luthra et al., 2019; Khandelwal & Barua, 2020; Gupta et al., 2021
Knowledge and skills among the workforce	KS	KS is the knowledge and expertise that employees must have to implement I4.0. New resource management skills among the workforce would help in establishing a clear connection between I4.0 and circular economy. Training in eco-friendly design, waste minimization skills, recycling, and reuse of materials are examples of skill sets and competencies	Cezarino et al., 2019; Luthra et al., 2019; Moktadir et al., 2020; Gupta et al., 2021
Data security and handling	DSC	I4.0 ensures that organizational data is secure and that it is used to improve operational efficiency. It allows for a seamless flow of information across the product lifecycle, from design to disposal	Liao et al., 2017; Patwa et al., 2020; Yadav et al., 2020
Adoption of Internet of Things technologies	IoT	IoT provides organizations with a technical infrastructure that allows them to collect and distribute real-time data about product/material return flows in the supply chain and manage waste using intelligent grids, sensors, and actuating devices	Rosa et al., 2020; Andrés et al., 2021; Gupta et al., 2021
Information sharing along value network	ISV	An advanced and efficient information-sharing system aids in the development of various activities such as shared resources; improving the process flow of materials or products along the value chain helps organizations to achieve better performance	Luthra et al., 2019; Rajput & Singh, 2019; Yadav et al., 2020

TABLE 12.1 (Continued)
List of identified drivers of Industry 4.0 (I4.0)

Drivers	Code	Description	References
Cloud-based design and manufacturing	CDM	The cloud-based solution ensures that the single source of items is maintained end to end, from suppliers to customers and vice versa. I4.0 and advanced manufacturing technologies for circular economy will be essential for sustainable organizations	Expert input
Waste recovery and management	WRM	I4.0 enables organizations to reduce losses by collecting more and more waste and recycling	Patwa et al., 2020; Garg, 2021; Salmenperä et al., 2021
Global standardization and sustainability goals	GSG	Global standards create a common language for all organizations to access their consistent effort in a sustainable direction. These standards enhance global compatibility and transparent operations	Yadav et al., 2020; Gupta et al., 2021
Long-term profitability	LTP	Circular economy reduces resource and material use by reusing, recycling, and remanufacturing, lowering the cost of product and increasing organizations' profitability	Horváth & Szabó, 2019; Khandelwal & Barua, 2020; Gupta et al., 2021
Health and safety	HS	Preparation of policies to monitor health and safety. Protection of workers from risks resulting from hazardous operations	Expert input

The steps involved in ISM methodology (Diabat & Govindan, 2011; Khandelwal & Barua, 2020) are as follows:

Step 1: Identify the drivers of I4.0 for CE based on the literature review; expert opinions were sought to consider prominent drivers for this study.
Step 2: Construct a contextual relationship matrix among the identified drivers based on expert opinion and brainstorming sessions.
Step 3: A pair-wise self-structural relationship matrix is established among the drivers considered.

Step 4: Develop a reachability matrix and check it for transitivity. The assumption for transitivity in ISM states that if criterion P influences criterion Q and criterion Q influences criterion R, then criterion P is necessarily influenced by criterion R.
Step 5: Partition the reachability matrix obtained in step 4 into different levels.
Step 6: Based on the levels obtained in step 5, the final ISM model is obtained.

12.3.1 Structural Self-Interaction Matrix (SSIM)

An SSIM has been established based on the contextual relationship among the drivers identified (Table 12.2). SSIM indicates the influence among the drivers that affect CE and I4.0 in the emerging markets.

The following symbols are used to give the direction of the relationship among the identified drivers to establish SSIM.

V: driver *i* will help in accomplishing driver *j*.
A: driver *j* will help in accomplishing driver *i*.
X: both driver *i* and *j* will help each other to accomplish.
O: *i* and *j* are not related.

12.3.2 Reachability Matrix

The reachability matrix is formed from the SSIM based on the following methods:

- If V is the entry in the SSIM matrix cell (*i*, *j*), 1 enters in the reachability matrix cell (*i*, *j*), and 0 enters the cell (*j*, *i*).
- If A is the entry in the SSIM matrix cell (*i*, *j*), 0 enters in the reachability matrix cell (*i*, *j*), and 1 enters in the cell (*j*, *i*).

TABLE 12.2
Self-structural interaction matrix

Enablers	HS	LTP	GSG	WRM	CDM	ISV	IoT	DSC	KS	MSC
MSC	O	O	O	O	V	O	V	V	O	–
KS	O	O	V	O	O	X	A	O	–	
DSC	O	O	O	V	X	O	O	–		
IoT	O	O	O	O	X	O	–			
ISV	O	O	O	X	A	–				
CDM	O	O	O	O	–					
WRM	O	V	O	–						
GSG	V	X	–							
LTP	V	–								
HS	–									

For abbreviations, see Table 12.1.

- If X is the entry in the SSIM matrix cell (i, j), 1 enters in both the cells (i, j) and (j, i) in the reachability matrix.
- If O is the entry in the SSIM matrix cell (i, j), 0 enters in both the cells (i, j) and (j, i) in the reachability matrix.

A final reachability matrix has been obtained after incorporation of the transitivity concept in the initial reachability matrix, as presented in Table 12.3. Transitivity states that if a driver P is related to Q and Q is related to R, then P and R are necessarily related.

12.3.3 Partition of Drivers in Levels

The final reachability matrix (Table 12.3) explains each identified driver's reachability, antecedent, and intersection sets. The drivers are divided into various levels based on their reachability and intersection sets. The reachability set of each driver consists of itself and the others which it will influence to achieve. The antecedent set of each driver consists of itself and the others which will influence to achieve it. The intersection set of each driver was also carried out. If the reachability and intersection set is the same for the given driver, this is considered to be level 1 and placed on the topmost position in the hierarchical structural model (Diabat & Govindan, 2011). Table 12.4 represents the first iteration. In the following iteration process, a driver in level 1 has been removed. The exact process has been carried out with the remaining drivers to get all the levels.

TABLE 12.3
Final reachability matrix

Enablers	MSC	KS	DSC	IoT	ISV	CDM	WRM	GSG	LTP	HS	Driving Power
MSC	1	1*	1	1	1*	1	1*	1*	1*	1*	10
KS	0	1	0	0	1*	0	1*	1	1*	1*	6
DSC	0	1*	1	1*	1*	1	1	1*	1*	1*	9
IoT	0	1	1*	1	1*	1	1*	1*	1*	1*	9
ISV	0	1	0	0	1	0	1	1*	1*	1*	6
CDM	0	1*	1	1	1	1	1*	1*	1*	1*	9
WRM	0	1*	0	0	1	0	1	1*	1*	1*	6
GSG	0	0	0	0	0	0	0	1	1	1	3
LTP	0	0	0	0	0	0	0	1	1	1	3
HS	0	0	0	0	0	0	0	0	0	1	1
Dependence power	1	7	4	4	7	4	7	9	9	10	62

For abbreviations, see Table 12.1.

TABLE 12.4
Level partition of drivers (iteration 1)

Drivers	Reachability set	Antecedent set	Intersection	Level
MSC	1 2 3 4 5 6 7 8 9 10	1	1	
KS	2 5 7 8 9 10	1 2 3 4 5 6 7	2 5 7	
DSC	2 3 4 5 6 7 8 9 10	1 3 4 6	3 4 6	
IoT	2 3 4 5 6 7 8 9 10	1 3 4 6	3 4 6	
ISV	2 5 7 8 9 10	1 2 3 4 5 6 7	2 5 7	
CDM	2 3 4 5 6 7 8 9 10	1 3 4 6	3 4 6	
WRM	2 5 7 8 9 10	1 2 3 4 5 6 7	2 5 7	
GSG	8 9 10	1 2 3 4 5 6 7 8 9	8 9	
LTP	8 9 10	1 2 3 4 5 6 7 8 9	8 9	
HS	10	10	10	I

For abbreviations, see Table 12.1.

12.3.4 Formation of ISM Model

See Figure 12.1.

12.3.5 MicMac Analysis

The MicMac analyses divided all the defined drivers into four clusters – autonomous, dependent, linkage, and independent – based on their driving and dependence strength. Table 12.3 presents the driving and dependence power of each driver. Drivers that have weak driving and strong dependence power come under the category of the dependent cluster. The drivers that have intense driving and dependence power fall under the category of linkage drivers. They require most attention due to their strong influence on other drivers. In the independent driver category, weak dependence power and strong driving power drivers fall under this cluster. Figure 12.2 represents the MicMac diagram of I4.0 drivers for CE initiatives.

12.4 RESULTS AND DISCUSSION

Through extensive literature review and various rounds of discussion with experts from industry and academia, ten drivers of I4.0 were considered in the current study for CE initiatives in the emerging markets. The ISM methodology was considered in this study to analyze the contextual relationship among the identified drivers. Figure 12.1 presents the ISM model, depicting that health and safety (10) drivers come in level 1 and obtain the topmost position in the five-level hierarchical ISM model. Global standardization and sustainability goals (8) and long-term profitability (9) are secured at level 2; knowledge and skillset among workforce (2), information sharing along value network (5), and waste recovery

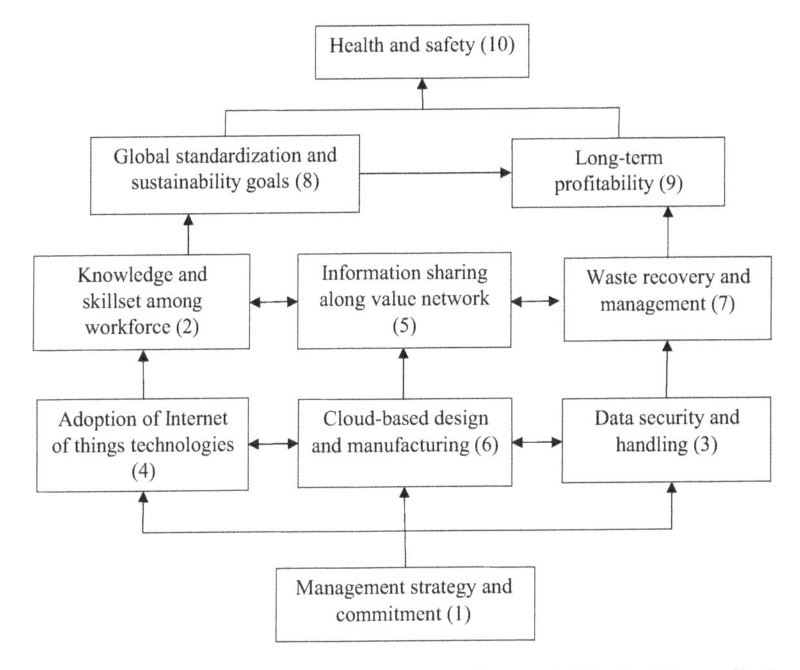

FIGURE 12.1 Interpretative structural modeling (ISM) model for the drivers of Industry 4.0 (I4.0) for circular economy (CE) initiatives.

and management (7) are secured at level 3; adoption of IoT technologies (4), cloud-based design and manufacturing (6), and data security and handling (3) are secured at level 4; and management strategy and commitment (1) obtained level 5 and is positioned at the bottom of the hierarchical model. Figure 12.2 presents the driving and dependence power diagram and summarizes all the identified drivers into four clusters. The following interpretation has been carried out from the MicMac diagram.

Autonomous Driver (Cluster 1): From Figure 12.2, no driver is present in this cluster with weak dependence and weak driving power. Consquently, all the I4.0 drivers considered in this study influence CE initiatives in the emerging markets.

Dependent Driver (Cluster 2): The drivers with low driving and high dependence power come under this cluster. In our study two drivers fall into this cluster, i.e., global standardization and sustainability goals, and long-term profitability having weak driving and strong dependence power. Both the drivers are strongly influenced and dependent on the other drivers considered in this study.

Linkage Driver (Cluster 3): Three drivers, namely knowledge and skillset among the workforce, information sharing along value network, and waste recovery and management having intense driving and dependence power, come under this cluster. These drivers, having strong influential and strong dependent powers and a small fraction change in one driver of this cluster, will quickly influence the remaining drivers under this cluster and overall influence system output.

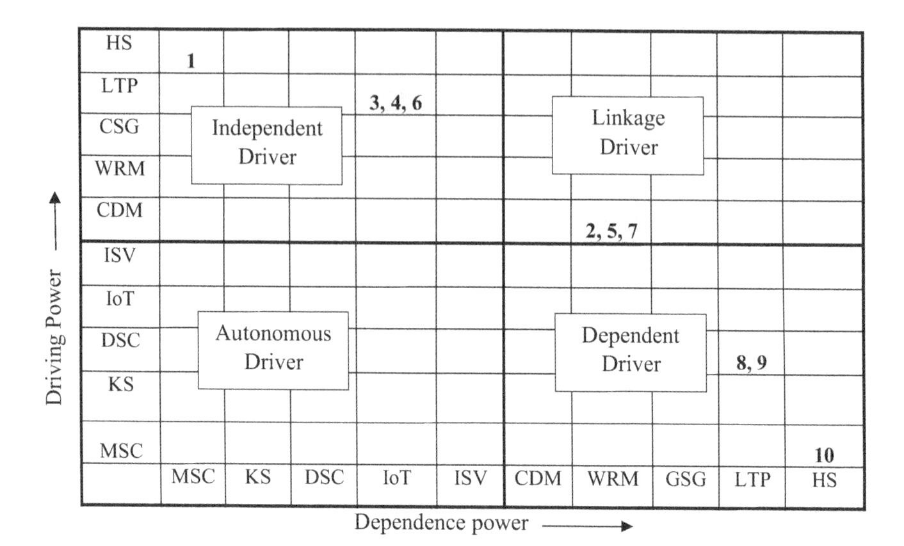

FIGURE 12.2 MicMac analysis for the drivers of Industry 4.0 (I4.0) for circular economy (CE) initiatives.

Independent Driver (Cluster 4): Four drivers come under this cluster, namely adoption of IoT technologies, cloud-based design and manufacturing, data security and handling, and management strategy and commitment. They have a solid capability to influence all other drivers. These drivers act as an essential key driver of I4.0 for CE initiatives. Management strategy and commitment, positioned at the bottom of the hierarchical model, suggests that management policies and commitment to implement such policies play an essential role in CE initiatives in I4.0. Top management decisions drive the organizations to move towards CE. Management strategies drive an organization to adopt IoT technologies, a cloud-based design framework in their supply chain processes, and data handling and management.

12.5 IMPLICATIONS OF THE STUDY

In the present study, we explored the drivers of I4.0 for the CE initiative in the emerging market. The current research highlights the following implications:

- This work helps managers and practitioners to understand the nature of the drivers (leading and lagging) of I4.0 for CE in the ISM model. The ISM model consists of driving variables such as management strategy and commitment. Top management of the organization must have proper plans and policies to implement the I4.0 technologies in the CE environment. Implementation of digital technologies brings transparency, trust, and security to the process.

- By implementing the IoTs in the streamlined operations, companies can get real-time data with location. That reduces costs by prioritizing and also reduces productivity congestion through process optimization. Managers of companies need to identify production and logistics operations to leverage the benefits of the IoT.
- Cloud allows the companies to store, manage, and process a large amount of data at minimum cost. Companies can save significant operating costs. Small company managers should purchase the cloud instead of setting up their own operating system.
- Managers have to adopt blockchain technology to secure data and bring transparency to production and logistics operations.
- The present study proposes the two new drivers of I4.0 for the CE (cloud-based design and manufacturing and health and safety) in the literature.
- The present research can be a benchmark for future studies to carry out more comprehensive analysis.

12.6 CONCLUSION

The CE is a closed-loop system that consists of reusing, repairing, refurbishing, and recycling existing material and products, whereas I4.0 improves productivity, efficiency, flexibility, and agility in the production and distribution process. The present research identifies the drivers and develops a structural model of I4.0 for the CE initiative in emerging markets. In the study, eight drivers were identified from the literature and two new drivers (cloud-based design and manufacturing and health and safety) from expert opinion. In the analysis, we developed ISM among the selected drivers. The hierarchical structure of the ISM model describes the leading and lagging drivers in the system. The results show that in the ISM model, management strategy and commitment is the leading driver, and health and safety are lagging drivers. Operation and supply chain managers need to focus on leading drivers. The company's top management should formulate policies and business plans to attain sustainability goals.

The present chapter has several limitations. First, in our study, we considered only ten drivers for ease of calculation. Further research could be done by increasing and changing the drivers of the study. Second, in our expert team, we considered 25 experts, with a background in the manufacturing and logistics sectors. Further research could be done by increasing the number of experts and their experience background. Lastly, we used the ISM method in our study; however, other multiple-criteria decision-making tools (TISM, AHP, technique for order of preference by similarity to ideal solution (TOPSIS), etc.) could further explore the research insights.

REFERENCES

Abdul-hamid, A., Helmi, M., Tseng, M., Lan, S., & Kumar, M. (2020). Impeding challenges on industry 4 . 0 in circular economy: Palm oil industry in Malaysia. *Computers and Operations Research*, *123*, 105052. https://doi.org/10.1016/j.cor.2020.105052

Andrés, C., Romero, T., Castro, D. F., Hamilton Ortiz, J., Khalaf, O. I., Vargas, M. A., Castro, C. A., Ortiz, D. F., & Khalaf, J. H. (2021). Synergy between circular economy and Industry 4.0: A literature review. *Sustainability (Switzerland)*, *13*. https://doi.org/10.3390/su13084331

Baaziz, A., & Quoniam, L. (2014). How to use big data technologies to optimize operations in upstream petroleum industry. *21st World Petroleum Congress*.

Bumblauskas, D., Mann, A., Dugan, B., & Rittmer, J. (2019). A blockchain use case in food distribution: Do you know where your food has been? *International Journal of Information Management*, *March*, 1–10. https://doi.org/10.1016/j.ijinfomgt.2019.09.004

Casado-Vara, R., Prieto, J., La Prieta, F. De, & Corchado, J. M. (2018). How blockchain improves the supply chain: Case study alimentary supply chain. *Procedia Computer Science*, *134*, 393–398. https://doi.org/10.1016/j.procs.2018.07.193

Cezarino, L. O., Liboni, L. B., Stefanelli, N. O., Oliveira, B. G., & Stocco, L. C. (2019). Diving into emerging economies bottleneck: Industry 4.0 and implications for circular economy. *Management Decision*. https://doi.org/10.1108/MD-10-2018-1084

Diabat, A., & Govindan, K. (2011). An analysis of the drivers affecting the implementation of green supply chain management. *Resources, Conservation and Recycling*, *55*(6), 659–667. https://doi.org/10.1016/j.resconrec.2010.12.002

Diabat, A., Kannan, D., & Mathiyazhagan, K. (2014). Analysis of enablers for implementation of sustainable supply chain management: A textile case. *Journal of Cleaner Production*, *83*, 391–403. https://doi.org/10.1016/j.jclepro.2014.06.081

Garg, C. P. (2021). Modeling the e-waste mitigation strategies using grey-theory and DEMATEL framework. *Journal of Cleaner Production*, *281*, 124035. https://doi.org/10.1016/j.jclepro.2020.124035

Gopal, P. R. C., & Thakkar, J. (2016). Analysing critical success factors to implement sustainable supply chain practices in Indian automobile industry: A case study. *Production Planning and Control*, *27*(12), 1005–1018. https://doi.org/10.1080/09537287.2016.1173247

Govindan, K., Palaniappan, M., Zhu, Q., & Kannan, D. (2012). Analysis of third party reverse logistics provider using interpretive structural modeling. *International Journal of Production Economics*, *140*(1), 204–211. https://doi.org/10.1016/j.ijpe.2012.01.043

Gupta, H., Kumar, A., & Wasan, P. (2021). Industry 4.0, cleaner production and circular economy: An integrative framework for evaluating ethical and sustainable business performance of manufacturing organizations. *Journal of Cleaner Production*, 126253. https://doi.org/10.1016/j.jclepro.2021.126253

Horváth, D., & Szabó, R. Z. (2019). Driving forces and barriers of Industry 4.0: Do multinational and small and medium-sized companies have equal opportunities? *Technological Forecasting and Social Change*, *146*, 119–132. https://doi.org/10.1016/j.techfore.2019.05.021

Kamble, S. S., Gunasekaran, A., & Sharma, R. (2018). Analysis of the driving and dependence power of barriers to adopt Industry 4.0 in Indian manufacturing industry. *Computers in Industry*, *101*(June), 107–119. https://doi.org/10.1016/j.compind.2018.06.004

Khandelwal, C., & Barua, M. K. (2020). Modelling the barriers to implement SSCM in Indian plastic manufacturing sector. *International Journal of Business Excellence*, *21*(4), 467–487. https://doi.org/10.1504/IJBEX.2020.108555

Kouhizadeh, M., Zhu, Q., & Sarkis, J. (2019). Blockchain and the circular economy: Potential tensions and critical reflections from practice. *Production Planning and Control*, *0*(0), 1–17. https://doi.org/10.1080/09537287.2019.1695925

Kumar, S., Raut, R. D., Nayal, K., Kraus, S., Yadav, V. S., & Narkhede, B. E. (2021). To identify Industry 4.0 and circular economy adoption barriers in the agriculture supply chain by using ISM-ANP. *Journal of Cleaner Production*, *293*, 126023. https://doi.org/10.1016/j.jclepro.2021.126023

Lele, A. (2019). Industry 4.0. *Smart Innovation, Systems and Technologies*, *132*, 205–215. https://doi.org/10.1007/978-981-13-3384-2_13

Liao, Y., Deschamps, F., De Freitas, E., Loures, R., Felipe, L., & Ramos, P. (2017). Past, present and future of Industry 4.0: A systematic literature review and research agenda proposal. International Journal of Production Research, 55(12). https://doi.org/10.1080/00207543.2017.1308576

Luiz Mattos Nascimento, D., Alencastro, V., Luiz Gonçalves Quelhas, O., Goyannes Gusmão Caiado, R., Arturo Garza-Reyes, J., Rocha-Lona Instituto Politécnico Nacional, L., Santo Tomás, E., City, M., & Tortorella, G. (2019). Exploring Industry 4.0 technologies to enable circular economy practices in a manufacturing context: A business model proposal. *Journal of Manufacturing Technology Management*, *30*(3), 607–627. https://doi.org/10.1108/JMTM-03-2018-0071

Luthra, S., Kumar, A., Kazimieras Zavadskas, E., Kumar Mangla, S., & Arturo Garza-Reyes, J. (2019). Industry 4.0 as an enabler of sustainability diffusion in supply chain: An analysis of influential strength of drivers in an emerging economy. International Journal of Production Research, 58(5). https://doi.org/10.1080/00207543.2019.1660828

Mangla, S. K., Luthra, S., Mishra, N., Singh, A., Rana, N. P., Dora, M., & Dwivedi, Y. (2018). Barriers to effective circular supply chain management in a developing country context. *Production Planning and Control*, *29*(6), 551–569. https://doi.org/10.1080/09537287.2018.1449265

Moktadir, M. A., Kumar, A., Ali, S. M., Paul, S. K., Sultana, R., & Rezaei, J. (2020). Critical success factors for a circular economy: Implications for business strategy and the environment. *Business Strategy and the Environment*, *29*(8), 3611–3635. https://doi.org/10.1002/bse.2600

Norta, A., Matulevĭcius, R., & Leiding, B. (2019). Safeguarding a formalized blockchain-enabled identity-authentication protocol by applying security risk-oriented patterns. *Computers and Security*, *86*, 253–269. https://doi.org/10.1016/j.cose.2019.05.017

Papadopoulos, T., Gunasekaran, A., Dubey, R., Altay, N., Childe, S. J., & Fosso-Wamba, S. (2017). The role of big data in explaining disaster resilience in supply chains for sustainability. *Journal of Cleaner Production*, *142*, 1108–1118. https://doi.org/10.1016/j.jclepro.2016.03.059

Parast, M. M., Adams, S. G., & Jones, E. C. (2011). Improving operational and business performance in the petroleum industry through quality management. *International Journal of Quality and Reliability Management*, *28*(4), 426–450. https://doi.org/10.1108/02656711111121825

Patwa, N., Sivarajah, U., Seetharaman, A., Sarkar, S., Maiti, K., & Hingorani, K. (2020). Towards a circular economy: An emerging economies context. *Journal of Business Research*, *May*, 1–11. https://doi.org/10.1016/j.jbusres.2020.05.015

Rajput, S., & Singh, S. P. (2018). Identifying Industry 4.0 IoT enablers by integrated PCA-ISM-DEMATEL approach. *Management Decision*. https://doi.org/10.1108/MD-04-2018-0378

Rajput, S., & Singh, S. P. (2019). Connecting circular economy and Industry 4.0. *International Journal of Information Management*, *49*(March), 98–113. https://doi.org/10.1016/j.ijinfomgt.2019.03.002

Rosa, P., Sassanelli, C., Urbinati, A., Chiaroni, D., & Terzi, S. (2020). Assessing relations between circular economy and Industry 4.0: A systematic literature review. *International Journal of Production Research*, *58*(6), 1662–1687. https://doi.org/10.1080/00207543.2019.1680896

Salmenperä, H., Pitkänen, K., Kautto, P., & Saikku, L. (2021). Critical factors for enhancing the circular economy in waste management. *Journal of Cleaner Production*, *280*, 124339. https://doi.org/10.1016/j.jclepro.2020.124339

Sousa, W. G. de, Melo, E. R. P. de, Bermejo, P. H. D. S., Farias, R. A. S., & Gomes, A. O. (2019). How and where is artificial intelligence in the public sector going? A literature review and research agenda. *Government Information Quarterly*, *36*(4), 101392. https://doi.org/10.1016/j.giq.2019.07.004

Tjahjono, B., Esplugues, C., Ares, E., & Pelaez, G. (2017). What does Industry 4.0 mean to supply chain? *Procedia Manufacturing*, *13*, 1175–1182. https://doi.org/10.1016/j.promfg.2017.09.191

Tseng, M., Tan, R. R., Chiu, A. S. F., Chien, C., & Chi, T. (2018). Circular economy meets Industry 4.0: Can big data drive industrial symbiosis? *Resources, Conservation & Recycling*, *131*(December 2017), 146–147. https://doi.org/10.1016/j.resconrec.2017.12.028

Xu, T., Baosheng, Z., Lianyong, F., Masri, M., & Honarvar, A. (2011). Economic impacts and challenges of China's petroleum industry: An input-output analysis. *Energy*, *36*(5), 2905–2911. https://doi.org/10.1016/j.energy.2011.02.033

Yadav, G., Kumar, A., Luthra, S., Garza-Reyes, J. A., Kumar, V., & Batista, L. (2020). A framework to achieve sustainability in manufacturing organisations of developing economies using Industry 4.0 technologies' enablers. *Computers in Industry*, *122*, 103280. https://doi.org/10.1016/j.compind.2020.103280

Zhou, X., Song, M., & Cui, L. (2020). Driving force for China' s economic development under Industry 4.0 and circular economy: Technological innovation or structural change? *Journal of Cleaner Production*, *271*, 122680. https://doi.org/10.1016/j.jclepro.2020.122680

13 Strategies to Manage Perishability in a Perishable Food Supply Chain

Anish Kumar,[1] Pradeep Kumar,[1] and Sachin Kumar Mangla[2,3]

[1] Department of Mechanical and Industrial Engineering, Indian Institute of Technology, Roorkee, Uttrakhand (UK), India

[2] Operations Management, Jindal Global Business School, O P Jindal Global University, Haryana, India

[3] Knowledge Management & Business Decision Making, Plymouth Business School, University of Plymouth, Plymouth, UK

13.1 INTRODUCTION

Supply chain management (SCM) has a substantial part to play in a firm's performance and has attracted serious research attention over the last few decades. Effectiveness and efficiency of a supply chain result in the strategic success of a company to a great extent. Production, marketing, distribution, warehouse management, supplier and retailer management are all activities in SCM. Raw material procurement, production, and final delivery of the finished product are various operations in a supply chain (Gandhi et al., 2015). Product characteristics have a direct or indirect impact these operations. One such characteristic is product perishability, which if not managed successfully can cause catastrophic failure in a perishable product supply chain (PPSCs).

A PPSC is a supply chain that has product perishability-related design considerations in it. Depending upon various supply chain conditions, the product decays, deteriorates over a period of time or becomes obsolete after a certain time. Such products are usually sensitive to environment and temperature and have limited shelf life. Thus, they require specific management strategies and incorporation of special design considerations into the supply chain. PPSCs require specific environment conditions – temperature, humidity, light exposure, etc. Depending on these conditions, varying perishability management strategies

DOI: 10.1201/9781003181644-13

may be adopted. Further, there is continuous product spoilage and degradation, as in the case of fruit and flowers, and in some cases the product value degrades after a certain time, e.g. newspaper, bread, milk. Appropriate supply chain decisions are required depending on how product value decreases over time (Blackburn and Scudder, 2009). A PPSC needs special consideration in terms of delivery timings, shelf life, inventory, product quality, storage and transportation conditions.

The issue of perishability is even more pressing when it comes to food products. Given that food is organic, it is highly susceptible to degradation through micro-organic growth, chemical reaction, moisture loss, freshness and quality loss. Most food products are inherently perishable, and require special considerations with regard to production, packaging, handling, storage and distribution for successful value delivery to the customer. Food being a necessary commodity, its quality is directly related to human health and survival. Thus, the issue of perishability is directly related to global food, and security. Thus, handling perishability is a core concern in food supply chains. The present study focuses on the operational perspectives of managing a perishable food supply chain (PFSC), and identifying different strategies to manage the perishability factor.

The paper is organized as follows. Section 13.2 presents a literature review of PFSC and discusses how perishability has been dealt with in recent literature. Section 13.3 presents a brief introduction to the theoretical aspects of the strategies identified in the PFSC literature to manage perishability. Section 13.4 presents the case of a dairy supply chain. Finally section 13.5 concludes the present study.

13.2 LITERATURE REVIEW

This section presents a literature review of recent research articles on the food supply chain with explicit focus on the perishability factor. The review uses a combined systematic literature review (SLR) and bibliometric analysis approach. Both SLR and bibliometric analysis are powerful literature review techniques. SLR provides a reliable, verifiable and repeatable methodology to systematically search, appraise and synthesize literature. We used the Scopus database to search literature, combining the search terms: "TITLE-ABS-KEY ("perishable" OR "perishability") AND TITLE-ABS-KEY ("food" OR "meat" OR "milk" OR "dairy" OR "beef" OR "vegetable" OR "fruit") AND TITLE-ABS-KEY ("supply chain")". The search resulted in 601 articles. However, to arrive at a literature list specifically focusing on our research problem we followed a filtration scheme. After excluding "open access" sources and including only journal articles written in English, we arrived at 275 articles. Further, to limit the scope of our research to operations management, we included only the subject areas "engineering", "decision sciences" and "business management and accounting". Thus, we arrived at 182 articles, which were further studied in detail to exclude articles that did not have a clear perishability-related study perspective. After qualitatively excluding articles that were not directly related to our research scope, we narrowed down to 98 articles, which were studied in detail to answer our research questions.

13.2.1 Descriptive Analytics

Understanding publication trends in recent years will certainly help to identifying new gaps in this domain. As previously mentioned, 98 articles were considered for this review; and plot of the annual number of articles is shown in Figure 13.1.

As can be seen, there has been an exponential increase in the number of publications on the PFSC. However, the increase in publication fluctuates, which shows that there is still a gap in the literature focusing explicitly on the perishability factor. But the increasing number of papers in PFSC considering perishability shows how important it is. Out of the total articles considered, 68% are from the last 5 years only. This trend is expected to grow steadily in coming years.

Authors with affiliations from the China, United Kingdom, United States and India contributed the most articles, with 14 articles from each of these nations, as shown in Figure 13.2. These were followed by authors with affiliations from Iran: 12 articles, Italy: 10 articles, France: 7 articles, Netherlands: 6 articles,

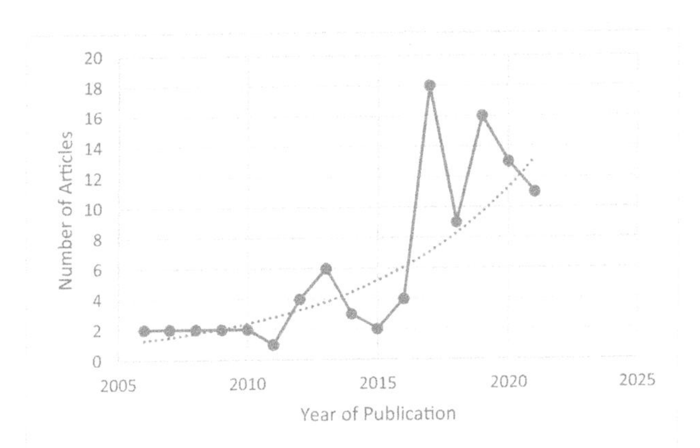

FIGURE 13.1 Number of publications by year.

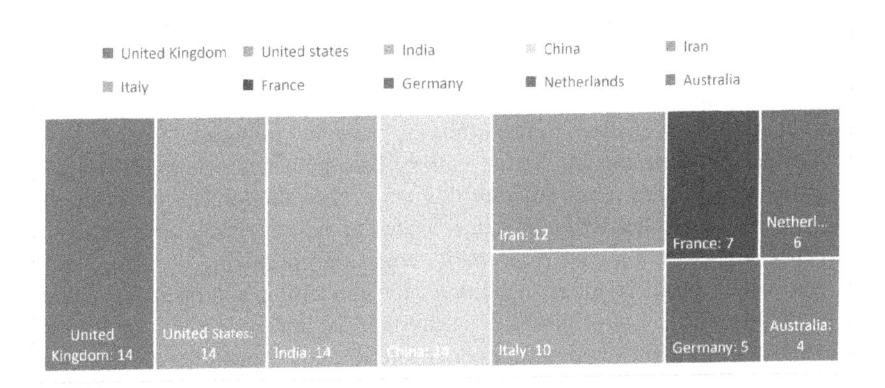

FIGURE 13.2 Number of publications by country.

Germany: 5 articles and Australia: 4 articles. As expected developed countries have contributed significantly. However, the increasing number of publications from developing countries certainly satisfies the call for more research there. Further, considering the leading publishing authors, only six authors had more than three articles. Dong Li was the author with the most articles, with six articles, followed by Xiaojun Wang and J.G.A.J. Van der Vorst, both with four articles. They were followed by Mehmet Soysal, Rene Haijema, Martin Grunow, with three articles each.

13.2.2 Bibliometric Analysis

For the articles that were finally selected, their relevant bibliometric details were downloaded in a ".csv" file format to conduct bibliometric analysis using the Vosviewer software. Bibliometric analysis was conducted using bibliographic coupling analysis, which relates articles based on the number of references they share. Those articles which share more references should have a similar research theme. Thus, studying software-generated clusters of articles from bibliographic coupling could provide us with an initial directive towards certain research themes, which would help us to answer our research questions. Further, analysis was conducted using co-occurrence analysis of text data in the abstract and title. This relates words based on the number of documents in which they co-occur; thus, words that co-occur more often should point towards a common direction.

The map generated from the bibliometric analysis is shown in Figure 13.3.

Figure 13.3 is a bibliometric map generated from co-occurrence analysis of the abstracts of the selected articles. We used binary counting of words with at least three occurrences. In total there were 252 such words. From these 252 words, we selected 100 words which were related to the perishability aspects of the PFSC. The co-occurrence map of these 100 words is presented in Figure 13.3. To avoid congestion, we have only shown links with strength of at least 5. Analysis of co-occurring word couples with strong linkage can significantly help in developing theoretical concepts. One example is shown in Figure 13.4.

As can be seen, the word "data" has strong linkages with seven words (quality, process, perishable product, time, performance, demand, and market). From this it can be inferred that data regarding processes, quality, time–temperature and market demand are crucial to manage PFSC. Similarly, other words and their linkages help us to form a conceptual understanding, presented later. Next, bibliographic coupling analysis of the articles is presented in Figure 13.5.

As can been seen, the articles are in three broad clusters, where articles with zero link strength are removed from this map. Thus the largest one is the red cluster with 37 articles. When the larger nodes in the red cluster are studied, we find that articles in this cluster primarily focus on optimizing the food supply chains while considering perishability in the modelling approach. To include perishability various strategies are used, such as time windows for delivery (Govindan et al., 2014), imposing product deterioration at each step of the supply chain network using arc multipliers (Yu and Nagurney, 2013), measuring quality deterioration for different storage time periods and temperatures (Zanoni and

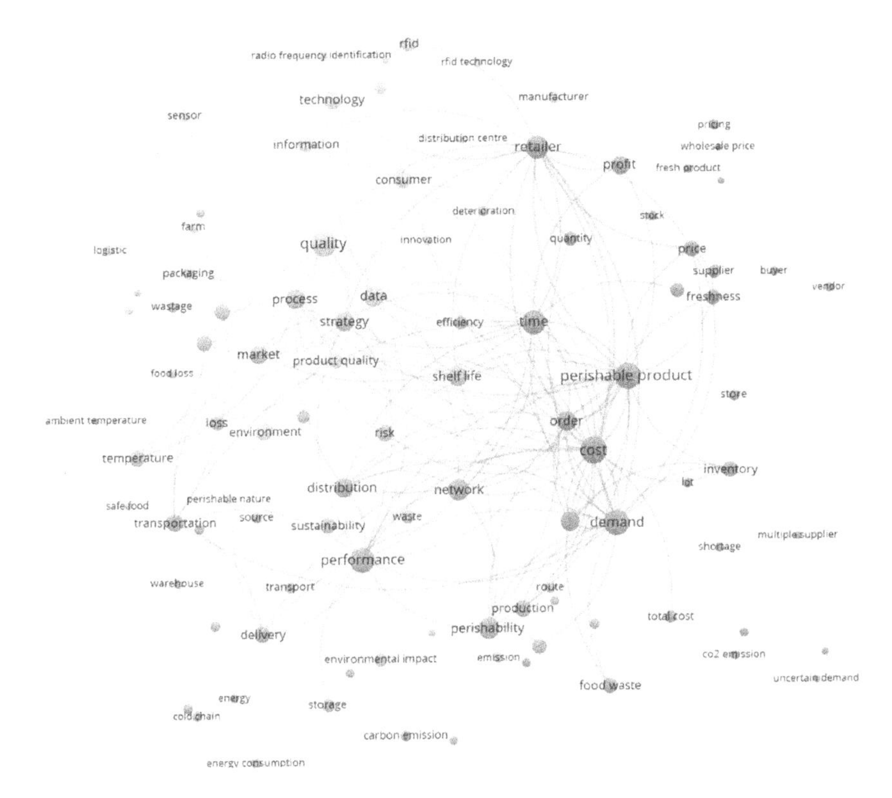

FIGURE 13.3 Co-occurrence analysis of the text from abstract.

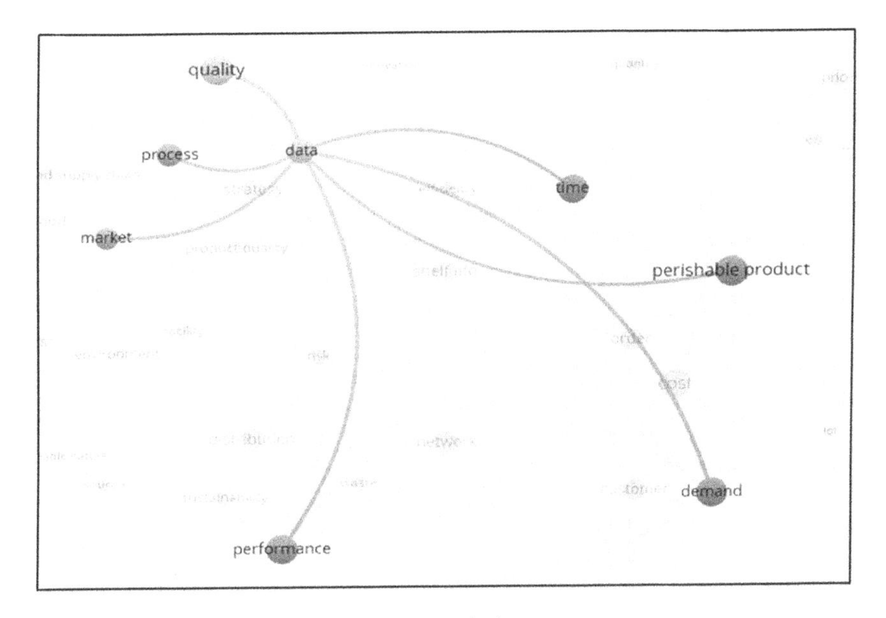

FIGURE 13.4 Example of co-occurrence analysis.

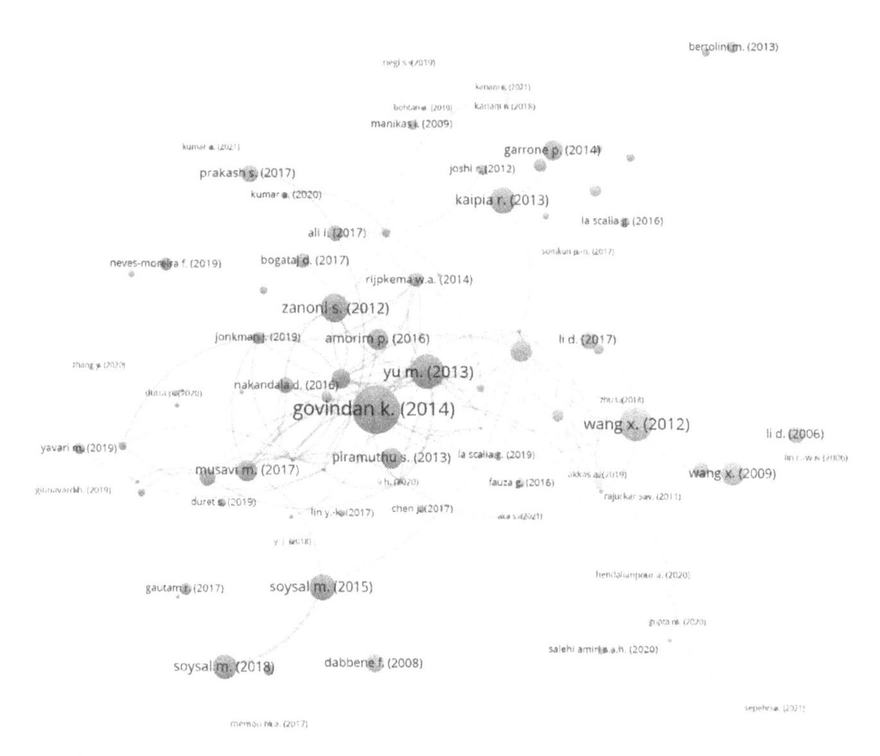

FIGURE 13.5 Bibliographic coupling analysis of the articles.

Zavanella, 2012), mapping heterogeneity of quality and variation of network structures (Keizer et al., 2017), uncertainty in food quality (W. S. De Amorim et al., 2018; Pedro Amorim et al., 2016) and as cost for waste. Thus, the first cluster represents the modelling and optimization approaches for PFSC including aspects of perishability.

In the second cluster, of 29 articles, study of its larger nodes reveals the primary focus is on perishability indicators and tracking and tracing of food quality, decay and shelf life. It suggests radiofrequency identification (RFID) and sensor-based technologies to predict decay and shelf life of perishable products (Grunow and Piramuthu 2013; La Scalia et al. 2016). The information gathered using such sensory technologies can be used to make better pricing decisions, reduce food wastage and improve product quality (Xiaojun Wang and Li, 2012; X. Wang, Li, and O'Brien, 2009).

The last cluster, which has 25 articles, includes articles which focus upon the operational perspectives of PFSC. Rijpkema, Rossi, and van der Vorst (2014) studied the sourcing strategies for perishable products and evaluated how different sourcing strategies could impact quality, shelf life and wastage in the PFSC. Gokarn and Kuthambalayan (2019) present a resources-based theoretical perspective of managing uncertainties in PFSC. Other authors in this cluster have also talked about supply chain resilience (Ali, Nagalingam, and Gurd, 2017), risk mitigation (Kumar et al., 2021), cold chain management (Joshi et al., 2012), procurement management (Kanani and Buvik, 2018), waste management (Kaipia, Dukovska-Popovska, and Loikkanen, 2013) and sustainability-related aspects of PFSC. As can be seen, there is a greater theoretical drive of the articles in this cluster; most use survey collection and emperical methodologies.

Thus we can see three clear aspects of research in PFSC, with those in the first cluster using optimization and mathematical modelling techniques, those in the second cluster principally focusing on technology application in PFSC and the last articles in the third cluster focusing on the theoretical aspects and operations management perspectives of PFSC. Thus, we have clear directions for research in PFSC identified through bibliometric analysis.

13.3 THEORETICAL BACKGROUND

Product perishability is an important factor that needs to be considered in a supply chain. It is an important factor to ensure safety, optimum quality and freshness of food products, limit wastages and simultaneously provide economic benefits. The inherent characteristic of food perishability is a risk factor in PFSC, resulting in concerns regarding safety for consumption, health of consumers, insufficient market access for producers and financial and environmental risks originating from food wastages. Thus, it becomes imperative that perishability is considered in supply chain planning; however, thisoften proves to be a difficult task. To manage perishability, the first and most frequently discussed issue is inventory management. Initial publications have primarily focused on inventory decisions related to PFSC. As the product is perishable, various inventory management policies, such as last-in-first-out (LIFO), first-in-first-out (FIFO), first-produced-first-out (FPFO), select-in-random-order (SIRO), collaborative planning, forecasting and replenishment (CPFR), can have different results. The inventory handling strategy of FIFO is generally suggested for perishable products. La Scalia, Micale, Miglietta, and Toma (2019) suggest first-expire-first-out (FEFO) as the most effective inventory management policy to synchronize residual shelf life of perishable products with the time required for transportation. Good results can be obtained through FPFO; however, it requires technology integration and efficient information dissemination to track production dates of individual products (Thron, Nagy, and Wassan, 2007). Strategies such as FPFO, FEFO and CPFR require significant technological interventions for inventory and shelf life monitoring purposes and information dissemination across the food supply chain. Thus, smart inventory management is a key strategy in PFSC to manage product perishability.

Bogataj, Bogataj, and Hudoklin (2017) suggested supporting FEFO inventory management by using cyber physical systems and internet of things infrastructure to measure ambient parameters to monitor changes in product shelf life in real time during logistics activity. RFID is another key technology that has been widely supported in the literature to manage PFSC. Effective deployment of RFID can provide real-time quality data of food products to constantly monitor food safety, quality and shelf life. Further, consumers are increasingly demanding quality, safety and sustainability-related information (Kumar, Mangla, Kumar, and Kayikci 2020), which necessitates the use of advanced technology. Such real-time quality data can help in tracing bad lots for easy product return, reduce quality risk and save liability costs by avoiding delivery as well as transportation of bad-quality product (Gautam et al., 2017). Thus, the cost of deployment of such a technology is over-ridden by savings in transportation, recall and returns, and reduced risk. Rossaint and Kreyenschmidt (2015) suggest using an intelligent labelling scheme with time–temperature indicators to manage PFSC. They propose there is up to 35% significant improvement in perishable food waste with time–temperature indicators.

CPFR is another key strategy to manage perishable inventory; it requires regular updating of store level sales and inventory data across the food supply chain. Vertical and horizontal collaborations are often discussed as key strategies in managing PFSC. Vertical collaboration is given more focus in supply chain literature, whereas horizontal collaboration is found to be of great importance in PFSC (Kumar, Mangla, Kumar, and Karamperidis, 2020). It improves the overall performance of the supply chain through improved resource utilization, such as refrigerated vehicles, cold storage, skilled labour, and so on. Improved resource utilization leads to reduced costs and emissions, and increased availability of resources and therefore reduced food wastage (Soysal et al., 2018).

Managing uncertainty related to demand, delays, process and supply inefficiencies is another critical aspect for the PFSC. Uncertainty related to product shelf life and quality is an inherent characteristic of PFSC. Researchers have used shelf life and quality-based pricing strategies to manage demand-related uncertainties as well as reduce wastages in PFSC. Aka and Akyüz (2021) consider uncertainties related to demand, employee performance and production cost to optimize inventory, waste and production costs. The current pandemic has brought uncertainties to the forefront. Uncertainties related to transportation, vehicle availability, routes, workforce availability and demand supply need to be considered while planning PFSC, which otherwise could create a huge amount of wastage and loss.

Stakeholder management is another important aspect of managing PFSC. Farmers, suppliers, third-party logistics (3PL) organizations, retailers, government organizations as well as customers play a critical role in ensuring quality, safety and perishability management in PFSC. A shared understanding of the pressing issues in PFSC and seamless information sharing among stakeholders is essential. Retailers play a key role in enforcing effective replenishment and ordering policies, product handling, minimizing temperature deviations, marketing and sale of new products. In PFSC enormous losses occur during the retail stage, and post-sale at the customer end. Thus sensitizing post-delivery stakeholders towards the issues of food waste and perishability is a key strategy in managing PFSC. Suppliers also play a key role as the uncertainty of the processes at the suppliers such as lead time and availability impact supply chain performance. Thus, many authors have focused on the issue of supplier selection in PFSC (Pedro Amorim et al., 2016). 3PL organizations ensure minimum delays in transportation and minimize product deterioration and loss in transportation.

Esteso, Alemany, and Ortiz (2021) have shown that including perishability-related constraints in supply chain design has significant economic benefits. Increasing product shelf life up to a certain extent was shown to have economic benefits with clear improvement from a food waste perspective. Thus, it is important to consider optimal product perishability characteristics to get optimal food supply chain design. Considering perishability is especially important from a logistics strategy perspective, as the longer a product spends in transportation or inventory, the greater is its loss of shelf life (P. Amorim and Almada-Lobo, 2014). To arrest loss of holding life and maximize shelf life, cold chain essentially forms the backbone of PFSC. However, while application of cold chain ensures extended holding life of perishables, it adds to supply chain costs and emissions. Thus, inventory holding costs, transportation costs and environmental emissions should be judiciously balanced with the economic and environmental benefits of reduced food waste through cold chain application in PFSC (Zanoni, Mazzoldi, and Ferretti, 2019). Refrigeration applications of warehouse, transportations and retail shops constitute more than 50% of the emissions of cold chains (Dong and Miller. 2021).

13.4 CASE STUDY

ABC Pvt. Ltd has been in the business of dairy products and fresh fruits and vegetables for more than three decades. It is a major player in the dairy market in the Delhi National Capital Region (NCR) of north India. It is an IS-15000 Hazard Analysis and Critical Control Point (HACCP), IS/ISO-9002 and IS-14001 environmental management systems (EMS)-certified organization. The company produces and markets a wide range of dairy products, such as milk, ice cream, culture products, butter, cheese, and so forth. It has a national production, sales and distribution network.

ABC conducts its operations through a network of private retailers, vendors, institutional customers and exclusive outlets. The present study specifically focuses on the supply chain of cultured products such as yogurt and buttermilk in the Delhi-NCR region. This division of their supply chain is called the cultured chain. Sale of these products also takes place through ABC exclusive outlets and retail sellers and through direct channels to bulk-buying parties. Transportation for the cultured chain is totally outsourced. The 3PL service providers are totally responsible for on-time delivery of product to the demand points. This case study evaluates how the organization manages perishability in this PFSC.

13.4.1 Managing the PFSC

The company functions on a farm-to-market cooperative model, with operations ranging from milk collection and processing to distribution and marketing. The operations are supported by cold chain throughout the supply chain. Being highly perishable, milk needs to be chilled immediately after production. The success of the supply chain is enabled by taking technology to the farm gate. Milk is collected twice a day from farmers at collection centres which are equipped with quality testing kits, chilling facilities and temporary storage facilities. From the collection points milk is sent to the nearest bulk chilling points. Here the milk is chilled to 2–4°C. This chilled milk is then transported to the processing plants. Thus, it follows the strategy for perishability management of first-mile cold chain application.

A great impetus is given to the quality of the initial produce. Thus, farmers are immediately paid at the collection point, but the payment is dependent upon the fat and solid non-fat content of the milk. The farmers are often trained to follow hygienic milking and storing techniques, and to ensure no delay in milking or transportation of milk. The farmers are also provided with support related to milch animal feed and health, which again ensure the quality of the milk. Thus, the most important strategy for managing perishability is supplier management. ABC has a range of suppliers, which range from individual producers to local dairy cooperatives. It has contracts with local dairies, which supply milk straight to ABC. These sublet organizations comply with the criteria set by ABC. They are also self-responsible for the quality of their product as well as activities like quality checks, packaging and transportation. In addition milk is sourced from individual producers, who are registered with a producer's institution, set up at village level by the organization. The producer's institution has collection units and weighing and testing facilities. The milk is then transported to the processing plant in insulated vehicles, processed as pasteurized milk or used as different milk products.

Transportation from distant locations takes place using insulated trucks as well as a daily train. The organization has four processing facilities in the Delhi-NCR region, the locations of which are strategically selected to maximize the reach and marketing capability of its products. The facilities are located to ensure a responsive supply chain; the facilities act as a centralized distribution facility to the proximity of the customer. This ensures more effective on-time deliveries. Thus, the strategic location of a processing facility is a key factor in managing PFSC.

Transporters should have the necessary infrastructure, expertise and manpower to provide a transportation service to ABC. ABC needs its product to be transported in an insulated vehicle from the plant to the retailers. The transporter will make available the required number of vehicles with insulated body as per requirements. The number of vehicles required by ABC is indicated from time to time with information supplied 1 day in advance.

ABC supplies its product of the requisite quality and quantity to the transporter against invoice from time to time based on the order received from concessionaires and retailers. The transporter then makes the supply of ABC milk to booths and other retail points as per the invoice and will make the delivery challan under its own signature, indicating the quality and quantity of various products supplied to and received by retailers as per the details of cheques / cash given by them.

The transporter, at its own cost, maintains, paints and ensures it meets the specifications and requirements of ABC. The transporter provides racks inside the insulated vehicles at its own cost according to the specifications and designs prescribed by ABC to accommodate the stocks of products. The transporter shall submit a security deposit as well In case of a rejection due to any kind of delay, the cost of product shall be recovered from the security deposited by the transporter. Thus, using outsourcing strategy activities ABC ensures the quality and performance of its distribution activities, which are critical for managing the perishability factor in PFSC. However, outsourced logistics are less reliable and there is less confidence in their services. The outsourced vehicles are frequently not maintained to the standards of the organization. General hygiene is also a significant concern in outsourced logistics. They lack ownership and commitment to the organization, thus may not perform. Thus, disciplinary issues also frequently occur.

Information flow serves as an important means of reducing complexity and improving the performance of a supply chain (Lusiantoro et al., 2018). The importance of information flow is greater in the PPSC. Given that product holding life and shelf life are limited, a lack of real-time sales and demand information can directly result in overstocking or understocking.

A customer would be less willing to buy an overstocked product that is approaching its expiry date. Further given that expiry dates are fixed but spoilage rates are stochastic, overstocked products are at greater danger of spoilage. Sale of a spoiled product is a direct cause of customer dissatisfaction. The demand for dairy products varies widely, due to festivals, marriages and climate, as well as day of the week. To maximize product availability, maintain customer loyalty and stay competitive in the market, it is important for ABC to have an efficient information flow.

ABC maintains a highly responsive demand management system. Procurement and sales are managed through SAP (System Application and Products) software. The company uses its on-plant inventory to balance demand and supply closely. Daily demand information is used to generate forecast and production schedules; it is exceptionally useful for milk and cultured products. Demand information is gathered through delivery channels as well as direct links with the sales end. ABC has a special call-in demand management centre that takes in demand from retailers and inputs demand of a product along with its the product code into the system. Complaints, sudden demand, bulk orders and queries are all handled through the demand management centre. Information from a warehouse management system generates real-time visibility of the inventory. Information management in ABC is done through ERP software that monitors inventory level, dispatch quantity, production level, delivered orders and pricing information. ABC maintains a highly accurate information management system to ensure high performance in its PFSC.

The retailer's end is the last leg of the PFSC, yet an essential one. Retailers perform the key functions of marketing products, managing stores, hiring retail staff, ordering supplies and providing responsive customer service. The company manages its retail stores on a franchise model. The retailers are thus not direct employees. However, ABC takes significant steps to train and integrate retailers as they have an important role in protecting the brand. The company supports retailers with responsive order processing and timely training, education and refresher courses. It provides quick complaint redressals in case of poor-quality product delivery, and also quality checks at the retail end.

The organization thus manages its perishable supply chain through effective and time-efficient logistics, effective cold chain operations, responsive inventory management, seamless information flow, use of outsourcing and 3PL services, effective stakeholder management and strict quality management. India being a tropical country, the climate is highly unsuitable for preservation of perishable food products like dairy products. Further, strict temperature control, highly responsive supply and procurement strategies, minimum time lag after milk procurement and quick reimbursement of farmers and producers, on-site quality checks at the producer's end have been the key distinguishing factors in ABC successfully managing perishability.

13.5 CONCLUSION

It is a challenge to manage perishability in the food sector. It causes significant economic and ecological loss through wastage of large amounts of food products globally. The present chapter sheds light on how best to manage this challenge in PFSC. The operational perspective of managing perishability is still not clear in the literature. The present chapter is an attempt to fill this gap. We have taken a mixed-methods approach where first a literature review was conducted using SLR and bibliometric analysis technique. The theory identified from the literature was backed with a case study analysis. We have tried to conceptually present the key aspects of managing the PFSC. The results of our study will be highly useful for academicians and industry personnel equally. However, much more research and analysis is required in this area. Future studies should focus on modelling and analysis of perishability-related factors, as well as providing decision support for managing this important aspect. Effective management of perishability across the food supply chain will contribute towards sustainable development goals, and is also important for global food safety and security.

REFERENCES

Aka, Salih, and Gökhan Akyüz. 2021. "An Inventory and Production Model with Fuzzy Parameters for the Food Sector." *Sustainable Production and Consumption* 26: 627–637. doi:10.1016/j.spc.2020.12.033.

Ali, Imran, Sev Nagalingam, and Bruce Gurd. 2017. "Building Resilience in SMEs of Perishable Product Supply Chains: Enablers, Barriers and Risks." *Production Planning and Control* 28 (15): 1236–1250. doi:10.1080/09537287.2017.1362487.

Amorim, P., and B. Almada-Lobo. 2014. "The Impact of Food Perishability Issues in the Vehicle Routing Problem." *Computers and Industrial Engineering* 67 (1): 223–233. doi:10.1016/j.cie.2013.11.006.

Amorim, Pedro, Eduardo Curcio, Bernardo Almada-Lobo, Ana P.F.D. Barbosa-Póvoa, and Ignacio E. Grossmann. 2016. "Supplier Selection in the Processed Food Industry under Uncertainty." *European Journal of Operational Research* 252 (3): 801–814. doi:10.1016/j.ejor.2016.02.005.

Blackburn, Joseph, and Gary Scudder. 2009. "Supply Chain Strategies for Perishable Products: The Case of Fresh Produce." *Production and Operations Management* 18 (2): 129–137. doi:10.1111/j.1937-5956.2009.01016.x.

Bogataj, David, Marija Bogataj, and Domen Hudoklin. 2017. "Mitigating Risks of Perishable Products in the Cyber-Physical Systems Based on the Extended MRP Model." *International Journal of Production Economics* 193 (June): 51–62. doi:10.1016/j.ijpe.2017.06.028.

De Amorim, Wellyngton Silva, Isabela Blasi Valduga, João Marcelo Pereira Ribeiro, Victoria Guazzelli Williamson, Grace Ellen Krauser, Mica Katrina Magtoto, and José Baltazar Salgueirinho Osório de Andrade Guerra. 2018. "The Nexus Between Water, Energy, and Food in the Context of the Global Risks: An Analysis of the Interactions Between Food, Water, and Energy Security." *Environmental Impact Assessment Review* 72 (March): 1–11. doi:10.1016/j.eiar.2018.05.002.

Dong, Yabin, and Shelie A. Miller. 2021. "Assessing the Lifecycle Greenhouse Gas (GHG) Emissions of Perishable Food Products Delivered by the Cold Chain in China." *Journal of Cleaner Production* 303: 126982. doi:10.1016/j.jclepro.2021.126982.

Esteso, Ana, M. M. E. Alemany, and Ángel Ortiz. 2021. "Impact of Product Perishability on Agri-Food Supply Chains Design." *Applied Mathematical Modelling* 96: 20–38. doi:10.1016/j.apm.2021.02.027.

Gandhi, Sumeet, Sachin Kumar Mangla, Pradeep Kumar, and Dinesh Kumar. 2015. "Evaluating Factors in Implementation of Successful Green Supply Chain Management Using DEMATEL: A Case Study." *International Strategic Management Review* 3. doi:10.1016/j.ism.2015.05.001.

Gautam, Rahul, Agnisha Singh, K. Karthik, S. Pandey, F. Scrimgeour, and M. K. Tiwari. 2017. "Traceability Using RFID and Its Formulation for a Kiwifruit Supply Chain." *Computers and Industrial Engineering* 103: 46–58. doi:10.1016/j.cie.2016.09.007.

Gokarn, Samir, and Thyagaraj S. Kuthambalayan. 2019. "Creating Sustainable Fresh Produce Supply Chains by Managing Uncertainties." *Journal of Cleaner Production* 207: 908–919. doi:10.1016/j.jclepro.2018.10.072.

Govindan, K., A. Jafarian, R. Khodaverdi, and K. Devika. 2014. "Two-Echelon Multiple-Vehicle Location-Routing Problem with Time Windows for Optimization of Sustainable Supply Chain Network of Perishable Food." *International Journal of Production Economics* 152: 9–28. doi:10.1016/j.ijpe.2013.12.028.

Grunow, Martin, and Selwyn Piramuthu. 2013. "RFID in Highly Perishable Food Supply Chains: Remaining Shelf Life to Supplant Expiry Date?" *International Journal of Production Economics* 146 (2): 717–727. doi:10.1016/j.ijpe.2013.08.028.

Joshi, Rohit, D. K. Banwet, Ravi Shankar, and Jimmy Gandhi. 2012. "Performance Improvement of Cold Chain in an Emerging Economy." *Production Planning and Control* 23 (10–11): 817–836. doi:10.1080/09537287.2011.642187.

Kaipia, Riikka, Iskra Dukovska-Popovska, and Lauri Loikkanen. 2013. "Creating Sustainable Fresh Food Supply Chains through Waste Reduction." *International Journal of Physical Distribution and Logistics Management* 43 (3): 262–276. doi:10.1108/IJPDLM-11-2011-0200.

Kanani, Renger, and Arnt Buvik. 2018. "The Effects of the Degree of Produce Perishability and the Choice of Procurement Channel on Supplier Opportunism: Empirical Evidence from the Food Processing Industry." *International Journal of Procurement Management* 11 (1): 113–133. doi:10.1504/IJPM.2018.088620.

Keizer, Marlies de, Renzo Akkerman, Martin Grunow, Jacqueline M. Bloemhof, Rene Haijema, and Jack G. A. J. van der Vorst. 2017. "Logistics Network Design for Perishable Products with Heterogeneous Quality Decay." *European Journal of Operational Research* 262 (2): 535–549. doi:10.1016/j.ejor.2017.03.049.

Kumar, Anish, Sachin Kumar Mangla, Pradeep Kumar, and Stavros Karamperidis. 2020. "Challenges in Perishable Food Supply Chains for Sustainability Management: A Developing Economy Perspective." *Business Strategy and the Environment* 29 (5): 1809–1831. doi:10.1002/bse.2470.

Kumar, Anish, Sachin Kumar Mangla, Pradeep Kumar, and Yaşanur Kayikci. 2020. "Investigating Enablers to Improve Transparency in Sustainable Food Supply Chain Using F-BWM." In *Intelligent and Fuzzy Techniques: Smart and Innovative Solutions, Proceedings of the INFUS 2020 Conference, Istanbul, Turkey, July 21–23, 2020*, pp. 567–575.

Kumar, Anish, Sachin Kumar Mangla, Pradeep Kumar, and Malin Song. 2021. "Mitigate Risks in Perishable Food Supply Chains: Learning from COVID-19." *Technological Forecasting and Social Change* 166 (January): 120643. doi:10.1016/j.techfore.2021.120643.

La Scalia, Giada, Luca Settanni, Rosa Micale, and Mario Enea. 2016. "Predictive Shelf Life Model Based on RF Technology for Improving the Management of Food Supply Chain: A Case Study." *International Journal of RF Technologies: Research and Applications* 7 (1): 31–42. doi:10.3233/RFT-150073.

La Scalia, Giada, Rosa Micale, Pier Paolo Miglietta, and Pierluigi Toma. 2019. "Reducing Waste and Ecological Impacts through a Sustainable and Efficient Management of Perishable Food Based on the Monte Carlo Simulation." *Ecological Indicators* 97 (October 2018): 363–371. doi:10.1016/j.ecolind.2018.10.041.

Lusiantoro, Luluk, Nicola Yates, Carlos Mena, and Liz Varga. 2018. "A Refined Framework of Information Sharing in Perishable Product Supply Chains." *International Journal of Physical Distribution & Logistics Management* 48 (3): 254–283. doi:10.1108/IJPDLM-08-2017-0250.

Rijpkema, Willem A., Roberto Rossi, and Jack G. A. J. van der Vorst. 2014. "Effective Sourcing Strategies for Perishable Product Supply Chains." *International Journal of Physical Distribution and Logistics Management* 44 (6): 494–510. doi:10.1108/IJPDLM-01-2013-0013.

Rossaint, Sonja, and Judith Kreyenschmidt. 2015. "Intelligent Label: A New Way to Support Food Waste Reduction." *Proceedings of Institution of Civil Engineers: Waste and Resource Management* 168 (2): 63–71. doi:10.1680/warm.13.00035.

Soysal, Mehmet, Jacqueline M. Bloemhof-Ruwaard, Rene Haijema, and Jack G. A. J. van der Vorst. 2018. "Modeling a Green Inventory Routing Problem for Perishable Products with Horizontal Collaboration." *Computers and Operations Research* 89: 168–182. doi:10.1016/j.cor.2016.02.003.

Thron, Thomas, Gábor Nagy, and Niaz Wassan. 2007. "Evaluating Alternative Supply Chain Structures for Perishable Products." *The International Journal of Logistics Management* 18 (3): 364–384. doi:10.1108/09574090710835110.

Wang, Xiaojun, and Dong Li. 2012. "A Dynamic Product Quality Evaluation Based Pricing Model for Perishable Food Supply Chains." *Omega* 40 (6): 906–917. doi:10.1016/j.omega.2012.02.001.

Wang, X., D. Li, and C. O'Brien. 2009. "Optimisation of Traceability and Operations Planning: An Integrated Model for Perishable Food Production." *International Journal of Production Research* 47 (11): 2865–2886. doi:10.1080/00207540701725075.

Yu, Min, and Anna Nagurney. 2013. "Competitive Food Supply Chain Networks with Application to Fresh Produce." *European Journal of Operational Research* 224 (2): 273–282. doi:10.1016/j.ejor.2012.07.033.

Zanoni, Simone, and Lucio Zavanella. 2012. "Chilled or Frozen? Decision Strategies for Sustainable Food Supply Chains." *International Journal of Production Economics* 140 (2): 731–736. doi:10.1016/j.ijpe.2011.04.028.

Zanoni, Simone, Laura Mazzoldi, and Ivan Ferretti. 2019. "Eco-Efficient Cold Chain Networks Design." *International Journal of Sustainable Engineering* 12 (5): 349–364. doi:10.1080/19397038.2018.1538268.

14 Six Sigma
Integration with Lean and Green

Dain D. Thomas,[1,2] *Dinesh Khanduja,*[3] *and Neeraj Kumar*[4]

[1] Assistant Professor, MRIIRS, Faridabad, India

[2] Research scholar, NIT-Kurukshetra, India

[3] Professor, NIT-Kurukshetra, India

[4] Assistant Professor, PKG College of Engineering and Technology, Harayana, India

14.1 INTRODUCTION

In the recent pandemic, customers' perceptions of the conventional view of quality have shifted towards products that create less waste and have less negative impact on the environment. These increasing environmental concerns of customers have forced organizations to rethink their business plans. Increased concern about sustainability and the environment has forced managers to plan fo the sustainable development of their organization without compromising on profit and efficiency.

In operation management, concepts such as green manufacturing focusing on the environment have impacts on production and operations without compromising on quality and cost. The main objective of such initiatives is to minimize resource use and waste and reduce pollution. In simpler terms, the objective is to improve the performance of the organization as regards its impact on the environment. Environmental concerns are multidimensional and can be considered under five main headings:

1. decrease in CO_2 emissions;
2. decrease in water and energy consumption;
3. decrease in business waste;
4. decrease in environmental waste;
5. increase in environmental cost.

Lean Six Sigma are the two of the most popular strategies and tools for process improvement aiming to maximize profit and stand in the competitive market. Integrating Lean and Six Sigma while taking into consideration the impact on the environment has led to organizations reducing waste, improving worker

DOI: 10.1201/9781003181644-14

commitment and continuous improvement. The integration of Lean with Six Sigma while keeping environmental considerations in mind has already become a familiar subject among academicians and in organizations dealing with both production and service sectors capable of improving sustainability and quality along with productivity, simultaneously reducing delays and waste.

This chapter will focus on the implementation of Lean Six Sigma (LSS) with consideration of the environment, also known as GLSS, discussing the framework and tools that can be used for implementation. The role of GLSS specifically in this integration will also be discussed.

14.2 DEFINITIONS OF LEAN AND SIX SIGMA

14.2.1 LEAN

Taiichi Ohno, an engineer working for Toyota, is the person behind the concepts of lean which explain the causes of waste produced and offer methods to reduce or if possible eliminate the amount of waste produced. In 1974, the economic problems faced by Japan meant that many industries experienced heavy losses in the market. However, Toyota continued to be a successful company even in such difficult economic times due to the Lean concept. Thanks to implementation of Lean methodology, the organization was not only able to manage its waste but also generated maximum value for the product.

The concepts of lean were first introduced in the manufacturing sector to increase output flexibility and transparency in the production system. Since then, it has been found ideal for the service sector, agencies and office as well as identifying waste generated within the system for any given process. Even though several types of waste are generated in a process, they have been broadly classified into eight categories:

1. defects: efforts caused by rework and scraps
2. overproduction: production that was not needed until demand rose
3. waiting: waiting time of people, information or system
4. poor utilization of talent: skilled worker not used to full potential
5. transport: unnecessary transportation of parts from one plant to another
6. inventory: extra materials or parts going unused
7. motion: unnecessary movement of parts within the work floor
8. extra processing: unnecessary processing of parts that do not add value.

Among the several waste reduction practices of lean the following have been widely used: 5S (sort, straighten, shine, standardize and sustain), value stream mapping (VSM), Kanban, Kaizen, Poka-yoke, single-minute exchange of dies (SMED), ISO 9000, JIT production system, total productive maintenance (TPM), total quality management (TQM) and Takt time. These practices have brought about the following business outcomes:

- improved quality
- reduced cost

- right time delivery
- customer satisfaction
- increased profits
- improved productivity
- reduced lead time.

14.2.1.1 Lean Integration with Green

The term green manufacturing was coined to refer to the innovative model that utilizes green strategies and techniques which focus on being eco-efficient. The major objectives of green manufacturing are to reduce negative environmental impacts and resource consumption in any given process. Green concerns itself majorly with pollution but in terms of operations management the major focus of green is the poor utilization of resources. The strategies that green uses to avoid poor use of resources and to reduce pollution are:

- product redesign
- process redesign
- 5R
- waste segregation
- ISO 14001.

These strategies all help in fulfilling the following outcomes of a green business:

- reduced pollution
- reduced materials use
- reduced cost
- increased productivity
- increased profitability
- optimum use of resources.

Growing interest in both strategies of lean and green has led to natural curiosity about their potential relationship. The business outcomes of both lean and green are very similar; both focus on poor use of resources in the form of defective products, scrap, unused inventory and over processing waste. If an industry adopts lean practices, then it automatically moves towards green and vice versa. Recent studies have suggested that industries can be green and reflect profits at the same time. Profits need not be sacrificed to environmental responsibility, or vice versa. So lean and green can be integrated and offered simultaneously, not only reducing environmental pollution but also providing optimum utilization of resources and improvement in productivity.

14.2.2 Six Sigma

The Six Sigma strategy was first created in the late 1980s at Motorola. It was intended to be a set of statistical techniques that managers use to measure the performance of a process. Using these techniques, a manager could bring changes to a process that would create improvements in efficiency. Once the process has achieved maximum efficiency, the managers then used Six Sigma statistical techniques to maintain process efficiency.

As Six Sigma gained popularity in the late 1990s, it was extended to improve processes to a great extent. The name Six Sigma was derived from theories of the standard bell-shaped curve. It is based on the fact that almost anything varies if you measure it with enough precision. For example, when machining a cylindrical rod to a diameter of 30 mm, if it is measured with low precision the rod will show 30-mm diameter, as desired. But with the help of high-precision measuring instruments, one rod may have the diameter specifications of 29.70 mm whereas another rod would measure 30.30 mm. Taking the average of both rods the average is 30.00 mm. But variations between rods do exist.

Statisticians described these forms of variations with a bell-shaped curve, also known as a Gaussian curve, named after the famous mathematician Carl Frederick Gauss, who was the first to work out the mathematics of variation during the early nineteenth century. The bell-shaped curve can be seen in Figure 14.1. Variations of frequently measured items follow the pattern of the bell-shaped curve.

In statistics, the Greek letter sigma (σ) denotes one standard deviation; however, the term sigma (σ) is different in context when referred to in statistics and TQM. It is essential to know the difference between σ used in statistics and the sigma applied in TQM. In statistics σ represents "value," of standard deviation from the mean, whereas in TQM, sigma represents a "level" where the upper and lower control limits are very significant. Hence, it is the kurtosis (kurtosis is the portion of data which represents whether it is heavy-tailed or light-tailed in relation to a normal distribution) of the control curve that decides the percentage of the values that lie within the two limits. The higher the kurtosis value, the higher the sigma level according to TQM concepts.

In TQM, the objective of Six Sigma is that if number of defects in a process are measurable then you can eliminate them and get as close to "zero defects" as possible. In order to achieve Six Sigma level of quality, a process must produce a maximum of 3.4 defects per million opportunities (DPMO). The term "opportunity" is defined here as the chance for non-conformance, or unable to meet

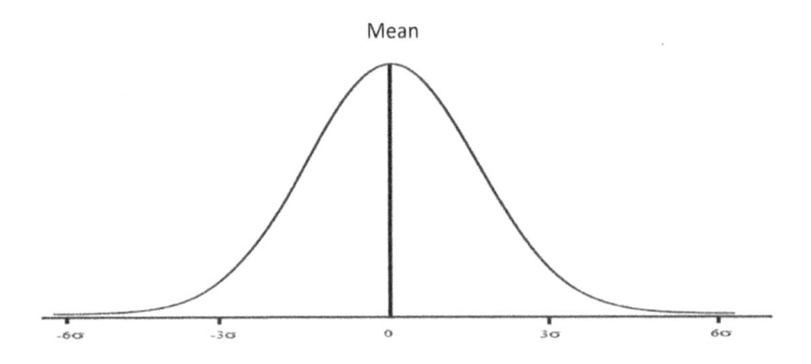

FIGURE 14.1 Bell-shaped or Gaussian curve with Six Sigma standard deviation.

TABLE 14.1
Defect and yield percentage of each sigma level

Sigma (σ) level	Defects per million opportunities (DPMO)	Percentage of defective (%)	Percentage of yield (%)
1 σ	691,462	69	31
2 σ	308,538	31	69
3 σ	66,807	6.7	93.3
4 σ	6210	0.62	99.38
5 σ	233	0.023	99.977
6 σ	3.4	0.0034	99.99966
7 σ	0.019	0.000019	99.9999981

the required specifications; therefore, the process needs to be nearly flawless. Table 14.1 shows the yield and defects percentage for each sigma level.

Six Sigma contains the following features:

1. customer perception of quality
2. defects that prevent delivery of what the customer wants
3. process capability
4. variation as seen and felt by the customer
5. consistent and predictable processes for improvement
6. designed to meet customer needs and process capability.

14.2.2.1 Definitions of Six Sigma

The previous section explained the concept of Six Sigma but what is the definition of Six Sigma? Several researchers and organizations have explained Six Sigma in several ways. A few of the prominent definitions are:

1. "Six Sigma is an operating philosophy that can be shared beneficially by everyone: customers, shareholders, employees, and suppliers. Fundamentally, it is also a customer-focused methodology that drives out waste, raises levels of quality, and improves the financial and time performance of organizations to breakthrough levels"(Joseph M. Juran).
2. "Six Sigma is a business management strategy that seeks to improve the quality of process outputs by identifying and removing the causes of defects (errors) and minimizing variability in manufacturing and business processes" (Wikipedia).
3. "Six Sigma is a data-driven method for achieving near perfect quality. Six Sigma analysis can focus on any element of production or service, and has a strong emphasis on statistical analysis in design, manufacturing, and customer-oriented activities" (UK Department for Trade and Industry).
4. "Six Sigma has three different levels:

- As a metric
- As a methodology
- As a management system

Basically, Six Sigma is all the three at the same time" (Motorola).

5. "Six Sigma is a highly disciplined process that helps us focus on developing and delivering near-perfect products and services. The word sigma is a statistical term that measures how far a given process deviates from perfection" (Generic Electric).

14.2.2.2 Six Sigma Integration with Lean

Over the years, Six Sigma has been found to be successful in several organizations including manufacturing, services, offices and agencies. Now several industries in Japan have successfully implemented Six Sigma, and are promoting it; even a few industries in India such as TVS Group are also moving towards Seven Sigma. With such prominence in both industry and academia, Six Sigma has found its way to integration with lean principles where process improvement would result in waste reduction.

14.3 METHODOLOGIES FOR SIX SIGMA

Basically, Six Sigma is a top-down approach practiced for management systems; It is is a high-performance system for executing business strategy at top management levels.

The methodology for its implementation has been effectively derived from Deming's popular *plan–do–check–act* (PDCA) cycle. The methodology of Six Sigma that has been generally used to improve a process or product, which is further advanced into the define, measure, analyze, improve and control (DMAIC) method. DMAIC is best used to improve the performance of existing processes or products. When improvement needs to start from the design phase (also known as design for quality) of a new product or process then, in order to achieve Six Sigma, Deming's PDCA cycle is further altered into define, measure, analyze, design and verify (DMADV). The methodologies for both Six Sigma systems of DMAIC and DMADV are shown in Figure 14.2.

The other Six Sigma methods related to DMAIC/DMADV that are commonly used are:

- IDOV (identify, design, optimize, validate)
- DCCDI (define, customer concept, design, implement)
- DMEDI (define, measure, explore, develop, implement).

Regardless of the name of the method, all of these methodologies use similar methods in quality function deployment: failure modes and effects analysis (FMEA), value engineering, simulation, robust design, design of experiments (DOE), methods improvement, statistical optimization, error proofing, etc. The only difference between them is if they are implemented to improve performance

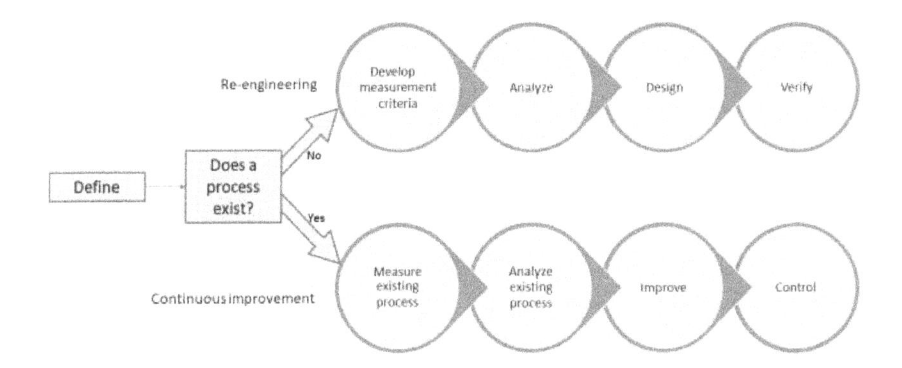

FIGURE 14.2 Working methodologies of define, measure, analyze, improve and control (DMAIC) and define, measure, analyze, design and verify (DMADV).

in an existing product or process (like DMAIC, methods improvement, etc.) or meant for new products or processes (like DMADV, value engineering, etc.). In DMADV, as in other acronyms, D can denote design in the case of new products or redesign in the case of existing products or processes.

14.3.1 DMAIC (Define, Measure, Analyze, Improve and Control)

In the previous section, the various methodologies of Six Sigma were discussed. Since the incorporation of Lean with Six Sigma is about improving an existing process in order to diminish or eliminate waste then DMAIC is the most suitable method [1, 2]. Hence this chapter will discuss DMAIC and its integration with lean, while keeping in mind environmental considerations. DMAIC stands for:

Define as clearly as you can customer requirements or problems.
Measure current performance levels and compare them using the voice of the customer.
Analyze the data collected regarding the existing process to determine the cause of the problem.
Improve: select a process design to solve the problem and then implement it.
Control the results and keep the gains of the selections made.

Six Sigma together with the DMAIC method has enabled several organizations around the globe to prosper and achieve higher levels of performance. Keen companies recognized that Six Sigma not only offers simple solutions to one-time problems, but implemented a novel system of doing business. Business challenges do not just go away in a free market; instead, they just keep changing and upgrading. Hence organizations felt comfortable in implementing Six Sigma with the DMAIC process. Figure 14.3 gives an overview of the process steps; although the time required for each step varies from process to process, the time periods for each step can overlap.

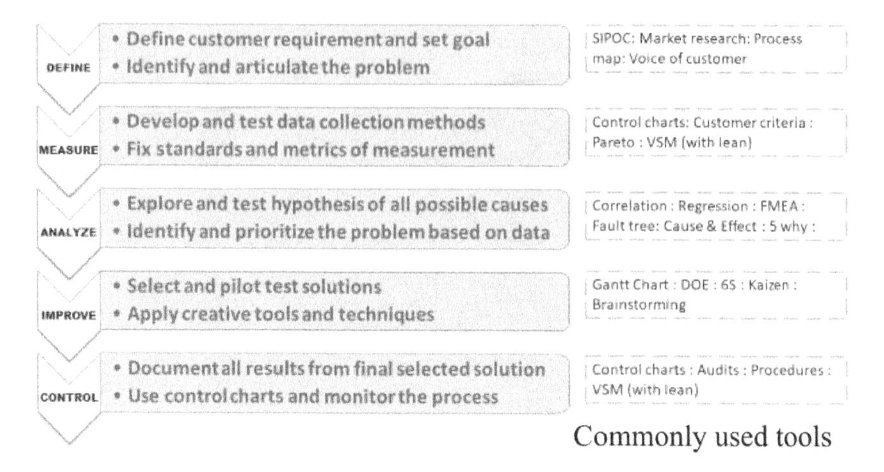

FIGURE 14.3 Overview of define, measure, analyze, improve and control (DMAIC) and commonly used tools and techniques. SIPOC, suppliers, inputs, process, outputs and customers; VSM, value stream mapping; FMEA, failure mode and effect analysis; DOE, design of experiments; 6S, Six Sigma.

The sequence of steps and the tools vary widely, depending on the size and complexity of the process or product for which improvement is required. In the best case, only one goal will be defined and measures will be created; obvious improvements are identified and implemented in process changes, then measures are created again. In the worst case, multiple goals will be identified and then the process is the same; measures are created for all goals, many possible improvements are identified. After improvements are attempted but acceptable results are not achieved, then the company must try again, selecting different measures for improvement, and then try again, analyze the measures, attempt another process improvement, and then measure some more, and lastly complete the revised goal. So in simple terms, a simple process can run straightforwardly, as described. But a complex project needs to be revised through the stages numerous times until the final goals have been implemented.

14.3.1.1 Define

The initial step of process improvement is very important. In the case of Six Sigma, in DMAIC methodology, the first step is to ask leaders and all involved to define the core processes. This can be used for an existing process or product, which does not meet customer expectations, and it is a very popular approach for Six Sigma process improvements. It is essential to define the scope, expectations, timelines and resources of a product, process or project. The definition of a project is a cooperative effort between the project manager and all those who have assigned the project, such as the process owners and project champions. This step of define identifies exactly what is part of the process or project and what is not, and describes the scope of the project, including the following:

- Find and define the scope of the problem.
- Recognize the customers and their requirements, through their voice. Define their requirements that are critical to quality.
- Define types of data, whether they are discrete or continuous, and identify data collection methods.
- Identify the project goals along with the project boundaries that define the above-mentioned scope.
- Identify a project charter, which is produced and approved during this step.

In the define phase of DMAIC a suppliers, inputs, process, outputs and customers (SIPOC) chart, voice of the customer, Kano model and critical-to-quality tree are the principal tools used, and some of these tools are also be used in other phases of DMAIC.

14.3.1.2 Measure

While the define phase focuses on knowing where you want to go, which is very important, the first information required before starting the journey is to know the current location, which is the second step of DMAIC: measure. In the measurement step the project manager calculates and creates a benchmark for an improvement opportunity, with the help of actual data. Six Sigma managers use specific tools to determine critical measures necessary to fulfill the customer's critical criteria by developing a measurement plan to document process/project/product performances.

The measure phase not only provides a goal, but also empowers managers to make comparisons after project completion, to establish whether significant improvement has been made and if so, how much. The measure phase is about relevant and substantial data, keeping the following in mind:

- Measure the gap between current and required performances.
- Gather the data in order to create a baseline for process performance capability for the output(s) of the process/project.
- Calculate a baseline performance and form a high-level process flow baseline.
- Assess the measurement system to be used, and check for adequate accuracy and precision.
- Calculate a current project sigma level.

The tools commonly used in this phase are process capability analysis, 5S, process map, Pareto, histograms and SIPOC chart, among others. Recently, the concept of a balanced scorecard was designed to help measure the actions of the organization's improvement strategy. A balanced scorecard is considered as a performance measurement system; it is derived from the project's vision and strategy, and hence reflects the critical features of the business.

14.3.1.3 Analyze

Once the project is defined and understood, and the baseline performance has been measured and documented, and it has been verified that there is an actual opportunity, then the next task is to do an analysis of the project/process. The determination of an analysis aims to discover the actual root cause of the problem among the several causes found in the data, and then validate and choose the best method for its elimination.

Several potential root causes have been recognized using root cause analysis tools and techniques, such as fish bone diagram (Ishikawa diagram), value analysis and method improvement studies which involve a fundamental breakdown. Statistical tests using *p*-values accompanied by Pareto charts/diagrams, histograms and line plots are also used for analysis. Then the top three to four potential root causes are selected with the help of team consensus tools such as multi-voting, which is further validated. The data is collected and analyzed in order to establish the contribution of each and every root cause that affects the project output(s). Then the process repeats itself unless and until a "valid" root cause is identified. After validation of the root causes:

- List and rank the potential causes of the problem.
- Arrange the root causes along with their process inputs; it is important to pursue this in the next step of DMAIC: improve.
- Validate all the causes and arrive at the root cause of the problem.
- Identify what sorts of effect these process inputs have on the process outputs. The collected data is analyzed to study the weight of contributions of each root cause, and then using the tools mentioned earlier, it is analyzed with the project metric.
- Generate thorough process maps to isolate the location of the root causes in the project/process, and factors contributing to the root cause occurrence.

Most Six Sigma projects will have a wealth of opportunity and alternatives, and it is during the analysis phase that the project managers or champions start to make decisions about what and when these can be accomplished.

14.3.1.4 Improve

In the previous step the project manager has identified and validated all root causes and the reasons for their occurrence. During the next step, which is the improve phase of DMAIC, all involved members have to propose ideas and solutions for these root causes, and implement them. The improvement phase requires project members to identify innovative methods to resolve customers' critical criteria defects, and finally to implement these innovative solutions. Another critical need in the implementation phase is that there must be regular checks to certify that the desired results are being achieved. Experiments and trials for these checks may be required in order to discover the best possible solution. This step largely revolves around the idea of creativity of the organization to resolve the problems and it can be ensured using the following steps:

1. Plan for a creative methodology.
2. Carry out brainstorming sessions with all involved members of the project.
3. Do value analysis and design of experiments for all possible solutions.

4. Test the solutions using the PDCA cycle.
5. FMEA is used to predict any avoidable risks associated with the process of improvement which have been achieved using the previously mentioned tools and techniques.
6. After all of the validations indicated above have been done, then create a complete implementation plan using Gantt charts and carry out the improvements.

Besides these steps, the concept of Kaizen has been found to be of great use in the improve phase, along with other tools such as benchmarking, Poke-yoke, 5S and simulations.

14.3.1.5 Control

After improvements have been implemented then it is essential in the Six Sigma project to make sure that performance-tracking mechanisms and measurements are in place to ensure, at least, that the profits and improvements made in the project are not lost over time and that there is continuous improvement. This requires the project manager to create a control plan and update documents, business processes and training records as and when required.

Another effective tool during the control stage is the control chart which evaluates the permanence of the desired improvements over a period of time by acting as a guide to continue observing and monitoring the project/process. The control charts can also provide a response plan for each of the improvements being examined in case the project/process becomes unstable.

Other steps used in the control stage are evaluating the results of process improvement, measurement systems analysis, standardization, mistake proofing, audit plan, total preventive maintenance and even FMEA.

In this stage it is encouraged that all the members (even outside the project) of the organization should be made aware of the implementations and all relevant documents should be shared for rapid implementation across the organization. This can lead to more ideas, feedback and innovative solutions to maintain the solutions from across the organization.

14.4 ORGANIZATION OF SIX SIGMA PROJECT MEMBERS

Six Sigma has been popular in several sectors in India and around the globe over the years [3–5] to maintain a high standard of quality. It is essential to specify those members of the team performing the tasks of Six Sigma to improve the quality of a project. Members who are to be included in a team are as follows:

1. *Executive leader*: everyone from the CEO to the top management members of the organization is responsible for the entire project and empower other organization members with the desired freedom and resources they need.
2. *Champions* take responsibility for Six Sigma application across the organization in a disciplined manner, while keeping in mind the considerations of top management. The executive leader identifies champions

from among upper management to mentor the Six Sigma black belts (see below).

3. *Master black belts* are recognized and handpicked by champions and work as full-time in-house coaches on Six Sigma. Their task is to support champions and guide black belts and green belts as they devote all their time to Six Sigma. They spend their entire time on assignments that ensure reliable use of Six Sigma across all departments within the organization.
4. *Black belts* are employed to work under master black belts and have to apply Six Sigma to specific assignments and projects, keeping the strategies of top management in mind. They are hired to focus full-time on execution of the Six Sigma project, whereas champions and master black belts are hired to identify projects and functions of Six Sigma.
5. *Green belts* are hired to personally implement Six Sigma in a project/process, while they work on other assignments simultaneously. They have to work under the leadership of black belts.
6. *Yellow belts* are educated in the simple application and implementation of Six Sigma management techniques and tools, as they work with all their seniors during all stages of project and are frequently the ones near the project location.

14.5 INTEGRATION OF GREEN, LEAN AND SIX SIGMA (GLSS)

In previous sections, principles of lean have been discussed as well as how integration of green and lean has been found to be integral in reducing waste and optimizing resources while keeping consideration of the environment. Six Sigma was discussed in detail along with the methodologies and organization of Six Sigma project teams.

14.5.1 Integration of Lean and Six Sigma

Over the years lean has been integrated with Six Sigma and found to be effective in reducing waste [6–9]. Academicians have researched the applications of LSS, such as Jiju Anthony [6], who compiled perspectives and viewpoints from academicians and practitioners having knowledge and experience in the fields of Lean and Six Sigma. It was concluded that both methods are focused on process and quality improvement. Lean focuses on speed and waste with experience in removing non-value-added entities, whereas Six Sigma focuses on variation, defects and process evaluation taking the organization to higher levels of performance.

As regards the principles of lean, others [10] have found that where the root cause of a problem is in a value-adding process or step of a project then Six Sigma has been very effective, as depicted in Figure 14.4.

Now the integration of Lean with Six Sigma (LSS) has been found to have the following benefits in any organization [11]:

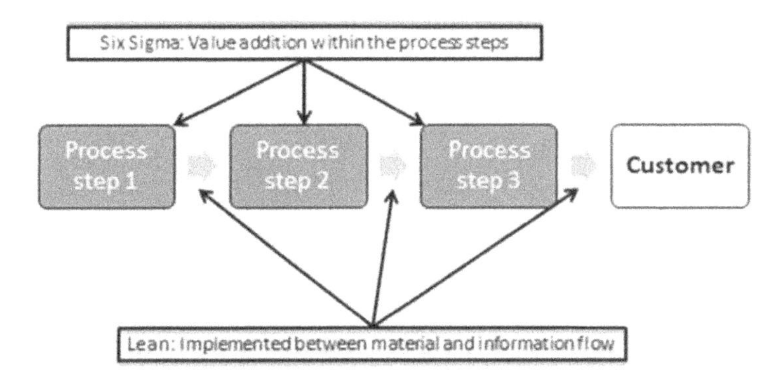

FIGURE 14.4 The Lean Six Sigma process view.

1. Costs have been reduced due to lack of direction in current problem-solving efforts.
2. The voice of the customer is associated with critical to quality (CTQ) and well understood.
3. Non-value-added operations can be drastically reduced.
4. LSS offers an important framework for managers to refer to.
5. There is good flexibility and responsiveness to meet customers' demands.

Several studies have been done on the integration of LSS, not just in the manufacturing sector [12] but also in healthcare and the public, service and education sectors [11].

14.5.2 Integration of Green Lean Six Sigma

The benefits of LSS have been expounded in earlier sections; the main objective of this chapter was integration of LSS in a project while keeping in mind environmental considerations, which are principles of green. The current economy is facing major setbacks due to the pandemic situation. Several studies have shown how degradation of our environment affected the pandemic. In a study by Harvard C. Chan (Center for Climate Health and Global Environment: www.hsph.harvard.edu/c-change/), it was noted that the rise in infections is due to climate change as animals that were not supposed to come in contact with other animals are creating opportunities for pathogens to come in contact with new species. In another study it was found that there is a higher rate of death due to coronavirus for people living in areas with higher air pollution. India is a country that currently has very low ambitions for environmental considerations; this is all the more reason why the pandemic has affected the nation deeply. So now the nation should look for existing methodologies in several sectors that can be integrated for the benefit of the environment. One such integration that can prove it has consideration for the environment for future generations is integration of LSS with green principles.

TABLE 14.2
Comparison of Green, Lean and Six Sigma

	Green	*Lean*	*Six Sigma*
Goal	Device environmentally friendly practices to improve environmental performances	Identify and reduce or eliminate waste	Increase the quality of projects of an organization by decreasing defects
Theories	Green design, manufacturing, planning and culture	Visual control, VSM, pull system, defect-free process	DMAIC
Comparison	Quality improvement by improvement in environmental practices	Quality improvement by reducing waste and non-value-added activities	Quality improvement by minimizing the defects in products and services
Applications	Supply chain, green product manufacturing and design sector	Public and service sectors, supply chain	Manufacturing industry, R&D organizations, service industry

VSM, value stream mapping; DMAIC, define, measure, analyze, improve and control; R&D, research and development.

The tools used in GLSS depend on the size of the project and type of organization – whether manufacturing or service – and several other factors. Although there have been fewer studies on GLSS, it is gaining in significance day by day across several sectors. Sanjay Kumar et al. [13] identified that the integration of GLSS is vital in product development as it helps to increase competitive advantage and achieve the sustainable objectives of the process. This author compared the three concepts of GLSS, as shown in Table 14.2.

There are a number of frameworks for implementation of GLSS for various industries in manufacturing and service and public sectors. In the next section conceptual frameworks across various sectors are discussed.

Some authors [14] have reflected on the fact that lean and green have certain limitations that can be reduced or eliminated using Six Sigma. They elaborated on DMAIC tools in integration with lean and green.

The authors discuss the *define* phase at a strategic level, which can help in prioritizing green and lean initiatives, defining which areas (facilities, marketing, logistics or manufacturing) of the organization require attention. The tools that are best suited are Pareto analysis, project ranking and selection matrix, project assessment matrix, Pareto priority index, quality function deployment, cost benefit analysis, multiple-criteria decision-making approaches such as analytic hierarchy process and theory of constraints.

For the *measure* phase reliable metrics of lean and green waste are measured and defined, such as extreme CO_2 emissions, energy and water consumption, poor utilization of raw material, inventory and transportation. By using lean tool VSM and green tool of life cycle assessment the source of waste can be established along with the scope of reduction.

In the *analyze* phase, the root cause of the problem can be determined along with all parameters involved. 5 whys and cause-and-effect diagrams can be used to find the cause of waste in water and energy. To get a scientific base of the root cause, popular tools include analysis of variance (ANOVA), scatter plots, DOE, hypothesis testing and regression analysis.

In the *improve* phase of DMAIC, tools and methods like 5S, brainstorming, poka-yoke and corrective action matrix can be used to eliminate the root cause of problems.

Finally, during the *control* phase, improvements are monitored with the help of statistical process control (SPC) to monitor consumption of water, raw materials, energy and CO_2 emissions.

14.5.3 Conceptual Frameworks of GLSS

Various frameworks have been designed for GLSS implementation; some are detailed below.

14.5.3.1 Automobile Sector

Sanjay Kumar [15], in further research, identified the enablers of GLSS for the Indian automobile sector. He developed a conceptual framework for the merger of the concepts of GLSS, keeping them on different levels of merger. Level 1 of this conceptual framework represents the simple concepts of their respective strategies, that are placed at level 2, whereas levels 3 and 4 discuss the possible mergers of GLSS, placed at level I and level II respectively.

Merger level I talks about LSS, GSS, Sustainable Green Six Sigma (SGSS), Green Lean (GL) and Sustainable Green Lean (SGL) as the opportunities for merging the concepts that have been drawn with the combined benefits of GLSS individually. Organizations which focus on improving environmental performances can lower their costs by reducing waste and may strive to export their products and services in global markets as well. Of all these mergers, in the last decade LSSmergers have been the most frequent.

Merger level II talks about the merger of GLSS and Sustainable Green Lean Six Sigma (SGLSS), which are not as prevalent yet claim to have been researched further. Several authors considered LSS as a very popular approach; in practice it seems to be environmentally friendly yet researchers have not entirely measured its environmental impact in the organization. Various mergers and levels need to be further researched to identify their benefits.

14.5.3.2 Jute Industry

Even in Bangladesh, GLSS has found its place in the jute industry, and has been researched by Talapatra and Gaine [16], who established a framework for the jute industry. The framework was built around DMAIC and integrated with lean tools and waste that can negatively impact the environment of a "precision spool" which keeps being rejected due to the inappropriate weight of the spool. The researchers understood the manufacturing of this spool and then implemented DMAIC with lean and green in mind. The tools used and results of each phase of this study are summarized in Table 14.3.

In terms of impact on the environment, electricity usage has been reduced by 7% and consumption of jute is reduced by 8%; this is after considering that rework is not required since defects have been reduced. Also, CO_2 emission was reduced by 72,128 kg/year due to decrease in energy consumption. In this study the framework was built in such a way that as the process was reducing defects it also analyzed the reduced impact on the environment.

14.5.3.3 Construction Sector

The construction sector in India is the second largest sector after agriculture and hence a major source of national consumption of resources. As previous studies have shown that GLSS has reduced consumption of resources across the globe for various sectors, then why not the construction sector? Hence researchers have successfully implemented GLSS in the construction sector [2, 17, 18].

One such study was done by Banawi and Bilec [17], who had the following objectives while designing the framework for GLSS:

TABLE 14.3
Define, measure, analyze, improve and control (DMAIC) tools and results in jute industry

Phase	Tool(s) used	From the phase
Define	SIPOC diagram, histogram	Minimum output of spools is 46 tons, out of which 14 tons did not meet the standards and was rejected
Measure	Process flow chart, life cycle assessment (LCA)	68% of spools were within the limits, hence the sigma level is 1. Usage of raw materials and energy consumption were also measured
Analyze	Pareto analysis, brainstorming, cause-and-effect diagram	The average thread count did not have a properly centered mean; the manufacturing process needed improvements
Improve	Statistical process control, 5S	Pilot studies for each problem cause showed process mean is now centered
Control	Control charts	Flow charts shared with all avoids jute piling and reduces wastage

SIPOC = suppliers, inputs, process, outputs and customers; LCA, 5S = sort, straighten, shine, standardize and sustain.

- Find waste at various points in the process using VSM: the lean tool.
- Measure the impacts on the environment of this waste using life cycle assessment (LCA).
- Remove or reduce the root causes of this waste using Six Sigma tools.

In this study the DMAIC tools consisted of tools from lean and green. For define and measure, VSM and LCA were used together to identify a process that was generating waste and how much. For analyze and improve, DMAIC tools of cause and effect and Pareto charts were used. For the control phase, re-evaluation of the process was done using VSM and LCA, again to check for any reduction in waste.

The authors used this framework on pile cap construction during a project. After implementing GLSS it was seen that consumption of materials had the highest impact on the environment, whereas a potential root cause of waste was identified as the regular design changes during the construction process. The framework will help contractors and managers to evaluate options and improve the efficiency of traditional methods.

14.6 CONCLUSION

GLSS are complementary; hence, each strategy has the potential to diminish each other's disadvantages. It is already know that lean is popular for its ability to identify waste generated in a process or project but it does not measure the impact on the environment. Green can assess the influence of the generated waste on the environment; hence, integrating lean and green together creates the ability to identify waste generated and measure the impact on the environment, although it is unable to provide a strategy to reduce waste by identifying its root cause. Six Sigma is the strategy that is able to reduce the waste generated by a process or project.

Across the globe, all sectors have implemented LSS, such as healthcare, services, manufacturing and even education, and a few sectors have successfully integrated GLSS. The automobile sector has used GLSS for product development and has a framework for merger of GLSS. In case of manufacturing and micro, small and medium enterprises (MSMEs), GLSS has found its place in the jute industry, which should encourage other MSMEs to practice tools and strategies that focus on the environment as well as improving quality and reducing waste. Another promising sector for GLSS implementation is the construction sector, which is the major consumer of resources in any country. Such strategies can reduce consumption of raw materials, water and energy, while at the same time reducing CO_2 emissions.

REFERENCES

1. A. J. Thomas, M. Francis, R. Fisher, and P. Byard, "Implementing Lean Six Sigma to overcome the production challenges in an aerospace company," *Prod. Plan. Control*, 2016, doi: 10.1080/09537287.2016.1165300.

2. M. Sony and S. Naik, "Green Lean Six Sigma implementation framework: a case of reducing graphite and dust pollution," *Int. J. Sustain. Eng.*, vol. 00, no. 00, pp. 1–10, 2019, doi: 10.1080/19397038.2019.1695015.
3. J. Antony and D. A. Desai, "Assessing the status of six sigma implementation in the Indian industry: results from an exploratory empirical study," *Manag. Res. News*, 2009, doi: 10.1108/01409170910952921.
4. K. Narasimhan, "The Six Sigma revolution: how General Electric and others Turned Process into Profits20021George Eckes. The Six Sigma Revolution: How General Electric and Others Turned Process into Profits . New York: John Wiley, 2001. 274 pp. (hardback), ISBN: IS," *TQM Mag.*, 2002, doi: 10.1108/tqmm.2002.14.1.68.1.
5. R. Rathi, D. Khanduja, and S. K. Sharma, "Capacity waste management at automotive industry in India: a Six Sigma observation," *Accounting*, January, pp. 109–116, 2016, doi: 10.5267/j.ac.2016.2.004.
6. J. Antony, "Six Sigma vs Lean: some perspectives from leading academics and practitioners," *Int. J. Product. Perform. Manag.*, vol. 60, no. 2, pp. 185–190, 2011, doi: 10.1108/17410401111101494.
7. R. J. Hilton and A. Sohal, "A conceptual model for the successful deployment of Lean Six Sigma," *Int. J. Qual. Reliab. Manag.*, vol. 29, no. 1, pp. 54–70, 2012, doi: 10.1108/02656711211190873.
8. N. F. Habidin and S. M. Yusof, "Critical success factors of Lean Six Sigma for the Malaysian automotive industry," *Int. J. Lean Six Sigma*, vol. 4, no. 1, pp. 60–82, 2013, doi: 10.1108/20401461311310526.
9. G. Yadav and T. N. Desai, "A fuzzy AHP approach to prioritize the barriers of integrated Lean Six Sigma," *Int. J. Qual. Reliab. Manag.*, vol. 34, no. 8, pp. 1167–1185, 2017, doi: 10.1108/IJQRM-01-2016-0010.
10. R. D. Snee and R. W. Hoerl, "Integrating Lean and Six Sigma: a holistic approach," *Six Sigma Forum Mag.*, vol. 6, no. 3, 2007.
11. J. Antony, R. Hoerl, and R. Snee, "An overview of Lean Six Sigma," *Lean Six Sigma High. Educ.*, vol. 34, no. 7, pp. 1–11, 2020, doi: 10.1108/978-1-78769-929-820201002.
12. M. Assarlind and L. Aaboen, "Forces affecting one Lean Six Sigma adoption process," *Int. J. Lean Six Sigma*, vol. 5, no. 3, pp. 324–340, 2014, doi: 10.1108/IJLSS-07-2013-0039.
13. S. Kumar, S. Luthra, K. Govindan, N. Kumar, and A. Haleem, "Barriers in green Lean Six Sigma product development process: an ISM approach," *Prod. Plan. Control*, vol. 27, no. 7–8, pp. 604–620, 2016, doi: 10.1080/09537287.2016.1165307.
14. J. A. Garza-Reyes, "Green lean and the need for Six Sigma," *Int. J. Lean Six Sigma*, 2015, doi: 10.1108/IJLSS-04-2014-0010.
15. S. Kumar, N. Kumar, and A. Haleem, "Conceptualisation of sustainable Green Lean Six Sigma: an empirical analysis," *Int. J. Bus. Excell.*, vol. 8, no. 2, pp. 210–250, 2015, doi: 10.1504/IJBEX.2015.068211.
16. S. Talapatra and A. Gaine, "Putting Green Lean Six Sigma framework into practice in a jute industry of Bangladesh: a case study," *Am. J. Ind. Bus. Manag.*, vol. 09, no. 12, pp. 2168–2189, 2019, doi: 10.4236/ajibm.2019.912144.
17. A. Banawi and M. M. Bilec, "A framework to improve construction processes: integrating Lean, Green and Six Sigma," *Int. J. Constr. Manag.*, vol. 14, no. 1, pp. 45–55, 2014, doi: 10.1080/15623599.2013.875266.
18. K. Hussain, Z. He, N. Ahmad, M. Iqbal, and S. M. Taskheer mumtaz, "Green, Lean, Six Sigma barriers at a glance: a case from the construction sector of Pakistan," *Build. Environ.*, vol. 161, June, 2019, doi: 10.1016/j.buildenv.2019.106225.

Index